Geometry and Computing

Contents

1   Introduction 1
   1.1   Two Fundamental Properties 1
   1.2   Different Possible Classifications 2
   1.3   Part I: Motivation 3
   1.4   Part II: Background – Metric and Measures 4
   1.5   Part III: Background – Polyhedra and Convex Subsets 4
   1.6   Part IV: Background – Classical Tools on Differential Geometry ... 5
   1.7   Part V: On Volume 6
   1.8   Part VI: The Steiner Formula 6
   1.9   Part VII: The Theory of Normal Cycles 7
   1.10 Part VIII: Applications to Curves and Surfaces 9

Part I   Motivations

2   Motivation: Curves 13
   2.1   The Length of a Curve 13
       2.1.1   The Length of a Segment and a Polygon 13
       2.1.2   The General Definition 14
       2.1.3   The Length of a C^1-Curve 15
       2.1.4   An Obvious Convergence Result 16
       2.1.5   Warning! Negative Results 16
   2.2   The Curvature of a Curve 17
       2.2.1   The Pointwise Curvature of a Curve 17
       2.2.2   The Global (or Total) Curvature 19
   2.3   The Gauss Map of a Curve 21
   2.4   Curves in $\mathbb{E}^2$ 22
       2.4.1   A Pointwise Convergence Result for Plane Curves 22
       2.4.2   Warning! A Negative Result on the Approximation
             by Conics 22
       2.4.3   The *Signed* Curvature of a Smooth Plane Curve 24

        2.4.4   The *Signed* Curvature of a Plane Polygon 26
        2.4.5   Signed Curvature and Topology 27
  2.5   Conclusion 28

3   Motivation: Surfaces 29
  3.1   The Area of a Surface 29
        3.1.1   The Area of a Piecewise Linear Surface 29
        3.1.2   The Area of a Smooth Surface 29
        3.1.3   Warning! The Lantern of Schwarz 30
  3.2   The Pointwise Gauss Curvature 33
        3.2.1   Background on the Curvatures of Surfaces 33
        3.2.2   Gauss Curvature and Geodesic Triangles 34
        3.2.3   The Angular Defect of a Vertex of a Polyhedron 36
        3.2.4   Warning! A Negative Result 37
        3.2.5   Warning! The Pointwise Gauss Curvature of a Closed
                 Surface 39
        3.2.6   Warning! A Negative Result Concerning
                 the Approximation by Quadrics 40
  3.3   The Gauss Map of a Surface 41
        3.3.1   The Gauss Map of a Smooth Surface 41
        3.3.2   The Gauss Map of a Polyhedron 42
  3.4   The Global Gauss Curvature 43
  3.5   The Volume 44

Part II Background: Metrics and Measures

4   Distance and Projection 47
  4.1   The Distance Function 47
  4.2   The Projection Map 49
  4.3   The Reach of a Subset 52
  4.4   The Voronoi Diagrams 55
  4.5   The Medial Axis of a Subset 55

5   Elements of Measure Theory 57
  5.1   Outer Measures and Measures 57
        5.1.1   Outer Measures 57
        5.1.2   Measures 58
        5.1.3   Outer Measures vs. Measures 58
        5.1.4   Signed Measures 59
        5.1.5   Borel Measures 60
  5.2   Measurable Functions and Their Integrals 60
        5.2.1   Measurable Functions 60
        5.2.2   Integral of Measurable Functions 61

5.3   The Standard *Lebesgue* Measure on $\mathbb{E}^N$ 62
        5.3.1   Lebesgue Outer Measure on $\mathbb{R}$ and $\mathbb{E}^N$ 63
        5.3.2   Lebesgue Measure on $\mathbb{R}$ and $\mathbb{E}^N$ 64
        5.3.3   Change of Variable 64
5.4   Hausdorff Measures 65
5.5   Area and Coarea Formula..................................... 66
5.6   Radon Measures 67
5.7   Convergence of Measures..................................... 67

Part III Background: Polyhedra and Convex Subsets

6   **Polyhedra** 71
6.1   Definitions and Properties of Polyhedra 71
6.2   Euler Characteristic 74
6.3   Gauss Curvature of a Polyhedron 75

7   **Convex Subsets** 77
7.1   Convex Subsets 77
        7.1.1   Definition and Basic Properties........................ 77
        7.1.2   The Support Function 79
        7.1.3   The Volume of Convex Bodies 80
7.2   Differential Properties of the Boundary 81
7.3   The Volume of the Boundary of a Convex Body 82
7.4   The Transversal Integral and the Hadwiger Theorem 84
        7.4.1   Notion of Valuation 84
        7.4.2   Transversal Integral 85
        7.4.3   The Hadwiger Theorem 86

Part IV Background: Classical Tools in Differential Geometry

8   **Differential Forms and Densities on $\mathbb{E}^N$** 91
8.1   Differential Forms and Their Integrals 91
        8.1.1   Differential Forms on $\mathbb{E}^N$ 91
        8.1.2   Integration of N-Differential Forms on $\mathbb{E}^N$ 93
8.2   Densities.. 94
        8.2.1   Notion of Density on $\mathbb{E}^N$ 94
        8.2.2   Integration of Densities on $\mathbb{E}^N$ and the Associated Measure . 95

9   **Measures on Manifolds** 97
9.1   Integration of Differential Forms............................... 97
9.2   Density and Measure on a Manifold 98
9.3   The Fubini Theorem on a Fiber Bundle 99

10   Background on Riemannian Geometry 101
    10.1 Riemannian Metric and Levi-Civita Connexion 101
    10.2 Properties of the Curvature Tensor 102
    10.3 Connexion Forms and Curvature Forms 103
    10.4 The Volume Form 103
    10.5 The Gauss–Bonnet Theorem 104
    10.6 Spheres and Balls 104
    10.7 The Grassmann Manifolds 105
        10.7.1 The Grassmann Manifold $G^o(N,k)$ 105
        10.7.2 The Grassmann Manifold $G(N,k)$ 106
        10.7.3 The Grassmann Manifolds $AG(N,k)$ and $AG^o(N,k)$ 107

11   Riemannian Submanifolds 109
    11.1 Some Generalities on (Smooth) Submanifolds 109
    11.2 The Volume of a Submanifold 112
    11.3 Hypersurfaces in $\mathbb{E}^N$ 113
        11.3.1 The Second Fundamental Form of a Hypersurface 113
        11.3.2 k^{th}-Mean Curvature of a Hypersurface 114
    11.4 Submanifolds in $\mathbb{E}^N$ of Any Codimension 115
        11.4.1 The Second Fundamental Form of a Submanifold 115
        11.4.2 k^{th}-Mean Curvatures in Large Codimension 116
        11.4.3 The Normal Connexion 116
        11.4.4 The Gauss–Codazzi–Ricci Equations 117
    11.5 The Gauss Map of a Submanifold 118
        11.5.1 The Gauss Map of a Hypersurface 118
        11.5.2 The Gauss Map of a Submanifold of Any Codimension 118

12   Currents 121
    12.1 Basic Definitions and Properties on Currents 121
    12.2 Rectifiable Currents 122
    12.3 Three Theorems 124

Part V   On Volume

13   Approximation of the Volume 129
    13.1 The General Framework 129
    13.2 A General Evaluation Theorem for the Volume 131
        13.2.1 Statement of the Main Result 131
        13.2.2 Proof of Theorem 38 131
    13.3 An Approximation Result 133
    13.4 A Convergence Theorem for the Volume 135
        13.4.1 The Framework 135
        13.4.2 Statement of the Theorem 137

14   Approximation of the Length of Curves 139
   14.1 A General Approximation Result 139
   14.2 An Approximation by a Polygonal Line 140

15   Approximation of the Area of Surfaces 143
   15.1 A General Approximation of the Area 143
   15.2 Triangulations 144
      15.2.1 Geometric Invariant Associated to a Triangle 144
      15.2.2 Geometric Invariant Associated to a Triangulation 145
   15.3 Relative Height of a Triangulation Inscribed in a Surface 145
   15.4 A Bound on the Deviation Angle 146
      15.4.1 Statement of the Result and Its Consequences 146
      15.4.2 Proof of Theorem 45 147
   15.5 Approximation of the Area of a Smooth Surface by the Area
         of a Triangulation 150

Part VI   The Steiner Formula

16   The Steiner Formula for Convex Subsets 153
   16.1 The Steiner Formula for Convex Bodies (1840) 153
   16.2 Examples: Segments, Discs, and Balls 155
   16.3 Convex Bodies in $\mathbb{E}^N$ Whose Boundary is a Polyhedron 158
   16.4 Convex Bodies with Smooth Boundary 159
   16.5 Evaluation of the Quermassintegrale by Means of Transversal
         Integrals 161
   16.6 Continuity of the Φ_k 162
   16.7 An Additivity Formula 164

17   Tubes Formula 165
   17.1 The Lipschitz–Killing Curvatures 165
   17.2 The Tubes Formula of Weyl (1939) 168
      17.2.1 The Volume of a Tube 168
      17.2.2 Intrinsic Character of the $\mathbb{M}_k$ 170
   17.3 The Euler Characteristic 171
   17.4 Partial Continuity of the Φ_k 171
   17.5 Transversal Integrals 172
   17.6 On the Differentiability of the Immersions 174

18   Subsets of Positive Reach 177
   18.1 Subsets of Positive Reach (Federer, 1958) 177
   18.2 The Steiner Formula 180
   18.3 Curvature Measures 182
   18.4 The Euler Characteristic 182
   18.5 The Problem of Continuity of the Φ_k 184
   18.6 The Transversal Integrals 186

Part VII The Theory of Normal Cycles

19 Invariant Forms 189
　19.1 Invariant Forms on $\mathbb{E}^N \times \mathbb{E}^N$ 189
　19.2 Invariant Differential Forms on $\mathbb{E}^N \times \mathbb{S}^{N-1}$ 190
　19.3 Examples in Low Dimensions 192

20 The Normal Cycle 193
　20.1 The Notion of a Normal Cycle 193
　　20.1.1 Normal Cycle of a Smooth Submanifold 194
　　20.1.2 Normal Cycle of a Subset of Positive Reach 194
　　20.1.3 Normal Cycle of a Polyhedron 195
　　20.1.4 Normal Cycle of a Subanalytic Set 196
　20.2 Existence and Uniqueness of the Normal Cycle 196
　20.3 A Convergence Theorem 198
　　20.3.1 Boundness of the Mass of Normal Cycles 199
　　20.3.2 Convergence of the Normal Cycles 199
　20.4 Approximation of Normal Cycles 200

21 Curvature Measures of Geometric Sets 205
　21.1 Definition of Curvatures 205
　　21.1.1 The Case of Smooth Submanifolds 206
　　21.1.2 The Case of Polyhedra 208
　21.2 Continuity of the $\mathcal{M}_k$ 209
　21.3 Curvature Measures of Geometric Sets 210
　21.4 Convergence and Approximation Theorems 210

22 Second Fundamental Measure 213
　22.1 A Vector-Valued Invariant Form 213
　22.2 Second Fundamental Measure Associated to a Geometric Set 214
　22.3 The Case of a Smooth Hypersurface 215
　22.4 The Case of a Polyhedron 216
　22.5 Convergence and Approximation 216
　22.6 An Example of Application 217

Part VIII Applications to Curves and Surfaces

23 Curvature Measures in $\mathbb{E}^2$ 221
　23.1 Invariant Forms of $\mathbb{E}^2 \times \mathbb{S}^1$ 221
　23.2 Bounded Domains in $\mathbb{E}^2$ 221
　　23.2.1 The Normal Cycle of a Bounded Domain 221
　　23.2.2 The Mass of the Normal Cycle of a Domain in $\mathbb{E}^2$ 223
　23.3 Plane Curves 224
　　23.3.1 The Normal Cycle of an (Embedded) Curve in $\mathbb{E}^2$ 224
　　23.3.2 The Mass of the Normal Cycle of a Curve in $\mathbb{E}^2$ 225

23.4 The Length of Plane Curves 226
    23.4.1 Smooth Curves 226
    23.4.2 Polygon Lines 227
23.5 The Curvature of Plane Curves 227
    23.5.1 Smooth Curves 227
    23.5.2 Polygon Lines 228

24 Curvature Measures in $\mathbb{E}^3$ 231
24.1 Invariant Forms of $\mathbb{E}^3 \times \mathbb{S}^2$ 231
24.2 Space Curves and Polygons 231
    24.2.1 The Normal Cycle of Space Curves 231
    24.2.2 The Length of Space Curves 232
    24.2.3 The Curvature of Space Curves 233
24.3 Surfaces and Bounded Domains in $\mathbb{E}^3$ 234
    24.3.1 The Normal Cycle of a Bounded Domain 234
    24.3.2 The Mass of the Normal Cycle of a Domain in $\mathbb{E}^3$ 235
    24.3.3 The Curvature Measures of a Domain 236
24.4 Second Fundamental Measure for Surfaces 238

25 Approximation of the Curvature of Curves 241
25.1 Curves in $\mathbb{E}^2$ 241
25.2 Curves in $\mathbb{E}^3$ 242

26 Approximation of the Curvatures of Surfaces 249
26.1 The General Approximation Result 249
26.2 Approximation by a Triangulation 250
    26.2.1 A Bound on the Mass of the Normal Cycle 250
    26.2.2 Approximation of the Curvatures 251
    26.2.3 Triangulations Closely Inscribed in a Surface 252

27 On Restricted Delaunay Triangulations 253
27.1 Delaunay Triangulation 253
    27.1.1 Main Definitions 253
    27.1.2 The Empty Ball Property 254
    27.1.3 Delaunay Triangulation Restricted to a Subset 255
27.2 Approximation Using a Delaunay Triangulation 256
    27.2.1 The Notion of ε-Sample 256
    27.2.2 A Bound on the Hausdorff Distance 256
    27.2.3 Convergence of the Normals 257
    27.2.4 Convergence of Length and Area 258
    27.2.5 Convergence of Curvatures 258

Bibliography 261

Index 265

Chapter 1
Introduction

The central object of this book is the *measure of geometric quantities* describing a subset of the Euclidean space $(\mathbb{E}^N, < ., . >)$, endowed with its standard scalar product.

Let us state precisely what we mean by a *geometric quantity*. Consider a subset $\mathcal{S}$ of points of the N-dimensional Euclidean space $\mathbb{E}^N$, endowed with its standard scalar product $< ., . >$. Let $\mathcal{G}_0$ be the group of rigid motions of $\mathbb{E}^N$. We say that a quantity $Q(\mathcal{S})$ associated to $\mathcal{S}$ is *geometric with respect to* $\mathcal{G}_0$ if the corresponding quantity $Q[g(\mathcal{S})]$ associated to $g(\mathcal{S})$ equals $Q(\mathcal{S})$, for all $g \in \mathcal{G}_0$. For instance, the diameter of $\mathcal{S}$ and the area of the convex hull of $\mathcal{S}$ are quantities geometric with respect to $\mathcal{G}_0$. But the distance from the origin O to the closest point of $\mathcal{S}$ is not, since it is not invariant under translations of $\mathcal{S}$. It is important to point out that the property of being geometric *depends on the chosen group*. For instance, if $\mathcal{G}_1$ is the group of projective transformations of $\mathbb{E}^N$, then the property of $\mathcal{S}$ being a circle is geometric for $\mathcal{G}_0$ but not for $\mathcal{G}_1$, while the property of being a conic or a straight line is geometric for both $\mathcal{G}_0$ and $\mathcal{G}_1$. This point of view may be generalized to any subset $\mathcal{S}$ of any vector space E endowed with a group $\mathcal{G}$ acting on it.

In this book, we only consider the group of rigid motions, which seems to be the simplest and the most useful one for our purpose. But it is clear that other interesting studies have been done in the past and will be done in the future, with different groups, such as the affine group (see [23, 36]), projective group, quaternionic group, etc.

1.1 Two Fundamental Properties

Our standpoint is that a geometric quantity is "interesting" if it possesses "fundamental" properties, related to the use one wants to make of it:

1. In applications like computer graphics, medical imaging, and structural geology for instance, scientists instinctively wish a *continuity condition*, related to the following simple observation: suppose one would like to evaluate a geometric

quantity $Q(S)$ defined on S, but one only has an approximation S' of S. It is natural to evaluate the quantity $Q(S')$, hoping that the result is "not too far" from $Q(S)$. In other words, one would like to write:

$$\text{if } \lim_{n\to\infty} S_n = S, \text{ then } \lim_{n\to\infty} Q(S_n) = Q(S). \qquad (1.1)$$

Note that this claim is incomplete since we have not specified the *topology* on the space $\mathcal{P}(\mathbb{E}^N)$ of subsets of $\mathbb{E}^N$. The simplest one is the Hausdorff topology, but we shall see that it is not enough in general.[1]

2. The second property is the *inclusion–exclusion principle*: basically, to evaluate a geometric quantity $Q(S)$ on a "big subset" S, it may be interesting to cut it into "small parts" S_i, evaluate $Q(S_i)$ on each "small part," and add the results to recover $Q(S)$. Roughly speaking, one wishes to have the equality:

$$Q(S_1 \cup S_2) = Q(S_1) + Q(S_2) - Q(S_1 \cap S_2). \qquad (1.2)$$

These two properties will be the "Ariane thread" of this book.

1.2 Different Possible Classifications

To classify such geometric quantities (as we have said, we only consider here quantities invariant under rigid motions), we can use topology or differential geometry:

- *Pointwise, local, or global geometric invariants.* A geometric property defined on S can be pointwise, local, or global. For instance:

    – The curvature of a smooth curve γ at a point, the Gauss curvature of a smooth surface of $\mathbb{E}^3$, and the solid angle of a vertex of a polyhedron are typical examples of pointwise properties. Although Fary [41] proved easily a convergence result for the curvature of a curve approximated by a sequence of inscribed polygons, the approximation of the pointwise curvatures of a surface approximated by a sequence of inscribed triangulations is very difficult (see [21]).
    – The area of a Borel subset of a surface S of $\mathbb{E}^3$ is a local invariant of S. Note that the continuity condition is not satisfied with the Hausdorff topology: the well-known Lantern of Schwarz [75] is a typical example of a sequence of polyhedra which "tends" to a smooth compact surface with finite area, although the sequence of areas of the polyhedra tends to infinity.
    – The genus of a closed surface of $\mathbb{E}^3$ is a global invariant. The Gauss–Bonnet theorem relates it to the Gauss curvature of the surface: integrating the Gauss curvature function over a closed surface gives its Euler characteristic (up to a constant).

---

[1] However, if one only considers the class of convex subsets, then a beautiful theorem of Hadwiger [56] states that the space of Hausdorff-continuous additive geometric quantities is spanned by the so-called *intrinsic volumes*. Examples of intrinsic volumes are length for curves, area for surfaces, but also integrals of mean and Gaussian curvature (for smooth convex sets).

- *Differential classification*. The minimum degree of differentials involved in the characterization of a geometric property on smooth objects may also be a way of classification. For instance:
  - The convexity property of an object $\mathcal{O}$ in $\mathbb{E}^N$ depends only on the position of the points of $\mathcal{O}$. No differential is involved.
  - The area of a (smooth) surface S involves only the first derivatives of a (local) smooth parametrization of S.
  - The curvature tensor of a Riemannian manifold M involves the second derivatives of a (local) smooth parametrization of M.

The aim of this book is to present a coherent framework for defining suitable *curvature measures* associated to a huge class of subsets of $\mathbb{E}^N$. These measures appear as local geometric invariants, involving 1 or 2 differentials in the smooth case (basically 1 for the length, the area, and the volume and 2 for the curvatures). These general geometric invariants coincide with the standard ones in the smooth case, but are also adapted to triangulations, meshes, algebraic and subanalytic sets, and "almost any" compact subsets of $\mathbb{E}^N$. Moreover, the continuity for a suitable topology and the inclusion–exclusion principle we mentioned at the beginning of this introduction will be systematically satisfied.

This book follows the long story of the sequence of little (and often brilliant) extensions of the classical notion of curvature. It appeared to the author that this historic presentation is also the most pedagogic approach to the problem.

We begin with the "old" geometric theory of convex subsets and end with "modern" computational geometry.

Let us now summarize the book, Part after Part, Chapter after Chapter:

- The motivations are made clear in Part I which deals only with curves and surfaces in $\mathbb{E}^3$.
- The essential material frequently involved in this book is given in Parts II–IV. It is a long background: it appeared important to provide the reader with the complete and precise material needed for the rest of the book. One needs topology, differential geometry, measure theory, and computational geometry. We summarize the results indispensable to the understanding of guiding ideas of the book.
- Parts V–VIII are the core of the book. They give the theory of the normal cycle, the definition of generalized curvatures of discrete objets, and convergence and approximation results. They end with an application to surfaces sampled by a finite cloud of points.

1.3 Part I: Motivation

Chapters 2 and 3 are an introduction to the subject, giving essentially simple examples and counterexamples to the problem of convergence of geometric quantities: the length of a smooth curve and its curvature in Chap. 2 and the area of a smooth surface and its mean and Gauss curvatures in Chap. 3. The problem of their discrete equivalents is introduced. We distinguish the pointwise vs. local or global versions of convergence.

1.4 Part II: Background – Metric and Measures

Chapter 4 deals with the distance map and the projection map in $\mathbb{E}^N$. It studies in detail their local and global properties. This leads us to recall the definition of the *reach* of a subset and the *Voronoi diagram* associated to a finite set:

1. The *reach* of a subset (also called the *local feature size*) was introduced by Federer [42]. The reach r of a subset S of $\mathbb{E}^N$ is defined as the maximal real number r such that the tubular neighborhood U_r of S of radius r has the following property: every point of U_r has a unique orthogonal projection on S. The real number r is always positive if S is smooth and compact. Of course, nonsmooth subsets may also have a positive reach. The main advantage of working in the class of subsets S with positive reach r is that the projection from U_r onto S is a smooth map, whose differential gives precise information on the shape of S.
2. On the other hand, the distance function allows one to define the *Voronoi diagrams* associated to finite sets, and more generally the *medial axis* of any subset of points.

Since the goal of this book is to build *curvature measures* on a large class of compact subsets of $\mathbb{E}^N$, Chap. 5 summarizes the basic and classical constructions of measures. It covers Lebesgue measure, the change of variable, and the area and coarea formulas.

1.5 Part III: Background – Polyhedra and Convex Subsets

The whole book deals with the approximation of smooth submanifolds by inscribed triangulations. That is why Chap. 6 is devoted to the indispensable background on polyhedra. It gives the main definitions on polyhedra and the precise definitions of the normal cone, the internal and external dihedral angles appearing in many explicit formulas of curvature measures. The chapter ends with the Gauss–Bonnet theorem for polyhedra.

Chapter 7 deals with convexity. The convex bodies are the first nonsmooth subsets on which global curvatures have been defined. For our purpose, two interesting properties of convex bodies are detailed:

1. The limit of a sequence of convex bodies is still convex.
2. The projection of a convex body on a hyperplane is still convex.

These two properties imply interesting results proved by induction on the dimension. This is the case for the Cauchy formula, which relates the volume of the boundary of a convex body with the integral of the volume of its projections on hyperplanes. The chapter ends with particular valuations on the class of convex bodies, and the Hadwiger theorem [52].

1.6 Part IV: Background – Classical Tools on Differential Geometry

Chapters 8 and 9 recall the definition of differential forms and densities on a manifold, and their relations with measures. In fact, we shall show later that particular differential forms integrated on a smooth submanifold give the classical mean curvature integrals of the submanifold. It will be the way to define *curvature measures* on a smooth object, and on any object on which such an integration can be done.

Chapters 10 and 11 give the necessary background on Riemannian geometry. The smooth objects studied in this book are submanifolds of $\mathbb{E}^N$, endowed with the induced metric and the Levi-Civita connexion. Chapter 10 introduces the intrinsic geometry of a Riemannian manifold, dealing with the curvature tensor. As examples, the spheres, the projective spaces, and the Grassmann manifolds are described. Chapter 11 deals with the extrinsic Riemannian geometry of submanifolds:

- We introduce the second fundamental form of any Riemannian submanifold, the principal curvature functions, and the kth-mean curvatures, generalizing the mean curvature and the Gauss curvature of surfaces. We are in particular interested in their integral over any Borel subset, which will appear in the second part of this book, in the tubes formula of Weyl. These integrals may be considered as *curvature measures*, which will be generalized to nonsmooth subsets, via the theory of normal cycles. The Gauss–Codazzi–Ricci equations, relating the extrinsic and intrinsic curvatures, are set out, since they will be used in technical proofs in the third part of the book.
- In classical theory of submanifolds, the Gauss map plays a key role. If M is a hypersurface of $\mathbb{E}^N$, the integral of the pullback by the Gauss map of the volume form of the unit hypersphere of $\mathbb{E}^N$ gives (up to a constant) the integral of the Gauss curvature of M, which is, by the Gauss–Bonnet theorem, its Euler characteristic. This result and its generalization in any dimension and codimension can be considered as the central point of the development of the theory of normal cycles.

Chapter 12 gives the basic background on currents, dual to differential forms. Indeed, currents can be considered as a generalization of submanifolds, on which differential forms can be integrated. We highlight the crucial *compactness theorem* for integral currents. It is the main tool in the proof of Fu's [48] convergence theorems for the curvature measures. We also mention a result on deformation of currents, used in the proof of the approximation results of the curvature measures.

1.7 Part V: On Volume

Part V covers the evaluation and approximation of the n-volume of a (measurable) subset of $\mathbb{E}^N$.

First of all, Chap. 13 studies deeply the well-known *Lantern of Schwarz* [75] (first example of a nonconvergence theorem for area). Then, it gives a general result: one can bound the n-volume of a smooth n-dimensional submanifold of $\mathbb{E}^N$ by the volume of another submanifold close to it, as far as one has information on their Hausdorff distance and their *deviation angle*, i.e., the maximum angle between their respective tangent bundles. It appears interesting to introduce a new geometric invariant, called *the relative curvature*, which connects the Hausdorff distance of the submanifolds and the second fundamental form of the initial one.

Chapter 14 applies the previous results to the evaluation of the length of curves approximated by a polygonal line, in terms of *the relative curvature*. The end of the chapter gives an useful bound of the deviation angle in terms of the length and the curvature of the curve.

Chapter 15 applies the previous results to surfaces in $\mathbb{E}^3$, approximated by triangulations. Another geometric invariant is introduced, namely the *relative height*, linking the length of the edges of the triangulation and the second fundamental form of the surface. With these new tools, elegant approximation and convergence theorems can be proven. They are corollaries of the general result stated in Chap. 13.

1.8 Part VI: The Steiner Formula

Part VI is concerned with the Steiner formula and its extensions.

Chapter 16 sets out the main theory of Steiner, discovered around 1840 (see [73] for instance). Given a convex body K of the Euclidean space $\mathbb{E}^N$, Steiner showed that the volume of the parallel body of K at distance ε is a polynomial of degree N in ε. When the boundary of K is smooth, the coefficients of this polynomial are, up to a constant depending on N, the integrals of the kth-*mean curvatures* of the boundary of K. Thus, these coefficients, called *Quermassintegrale* by Minkowski, are good candidates to generalize curvatures to convex hypersurfaces, without assuming any regularity condition. The problem of *continuity* of curvatures first appeared in this context: it could be proven that, if a sequence of convex bodies K_n has a Hausdorff limit K, then the *curvatures* of K_n converge to those of K. Using integral-geometric considerations, tight estimates can even be obtained for the difference between the *curvatures* of K_n and those of K.

Chapter 17 sets out the extension by Weyl [82] in 1939 of the results of Steiner, namely the *tubes formula*: Weyl proved that the interpretation of integrals of curvatures in terms of the volume of parallel bodies also holds if one drops the convexity assumption but assumes smoothness, provided ε is small enough. However, continuity with respect to the Hausdorff topology does not hold for smooth submanifolds, unless one assumes additionally that the curvatures of the sequence of submanifolds are uniformly bounded from above [42]. Under these assumptions, estimates of differences of curvature measures are known.

Chapter 18 sets out a part of the deep work of Federer [42] on geometric measure related to *curvature measures* (published in 1959). This author made a breakthrough in two directions:

1. He defined a large class of subsets, including smooth submanifolds and convex bodies, for which it is possible to define reasonable generalizations of curvatures: the *subsets of positive reach*. His approach consists again in considering the volume of parallel bodies. Basically, he observed that the key point in the tubes formula for both smooth and convex cases is that the orthogonal projection on the studied subset is well defined in a neighborhood of it. Subsets of positive reach are defined to be those for which this property holds.
2. He showed that one can actually associate to each subset K with positive reach in $\mathbb{E}^N$ and each integer $k \leq N$ a *measure* on $\mathbb{E}^N$, called the kth-*curvature measure* of K. When K is a smooth submanifold, its kth-curvature measure evaluated on a Borel subset U is nothing but the integral of the kth-mean curvature of K on U. *Curvature measures* thus give a much finer information than the *Quermassintegrale* since they determine, in the smooth case, the kth-mean curvatures at any neighborhood of any point of the subset.

Continuity with respect to the Hausdorff topology still holds for subsets with positive reach, if one assumes additionally a boundness condition on the reaches [42].

1.9 Part VII: The Theory of Normal Cycles

Unfortunately, Federer's approach could not handle some simple objects such as nonconvex polyhedra. Part VII is devoted to the theory of normal cycles, whose goal is to extend the results of Federer to subsets more general than subsets with positive reach. This step has been accomplished by Wintgen [83] in 1982 and Zähle [87]. These authors noticed that, in the smooth case, curvature measures of a smooth submanifold M of $\mathbb{E}^N$ arise as integrals over the unit normal bundle $ST^{\perp}M$ of the pullback of $(N-1)$-*differential forms* defined on the unit tangent bundle $ST\mathbb{E}^N$ of $\mathbb{E}^N$, which are invariant under rigid motions. In other words, the geometry of a submanifold is thus contained in the current determined by its unit normal bundle, by attaching to it a basis of the space of differential $(N-1)$-forms "invariant under rigid motions."

That is why Chap. 19 classifies these differential forms "invariant under rigid motions," defined on the unit tangent bundle of $\mathbb{E}^N$. It appears that a basis of this space can be simply and explicitly described.

Since singular spaces do not have in general a smooth normal bundle on which these invariant forms can be integrated, the main point is now to introduce a generalization of the normal bundle of a smooth object. The choice of Wintgen [83] is quite natural: using the duality between differential forms and currents, he introduced the concept of a *normal cycle* associated to a singular space.

Chapter 20 gives the details of this generalization. Associated to ("almost any") compact subset $\mathcal{A}$ of $\mathbb{E}^N$, Wintgen defined a closed integral current $\mathbf{N}(\mathcal{A})$, called the *normal cycle* associated to $\mathcal{A}$. An important property of the normal cycle is its additivity (which we also call the *inclusion–exclusion principle*): if $\mathcal{A}'$ is another compact subset of $\mathbb{E}^N$, one has

$$\mathbf{N}(\mathcal{A} \cup \mathcal{A}') = \mathbf{N}(\mathcal{A}) + \mathbf{N}(\mathcal{A}') - \mathbf{N}(\mathcal{A} \cap \mathcal{A}') \qquad (1.3)$$

whenever both sides are defined. In particular, the normal cycle of a not necessarily convex polyhedron can be computed from any triangulation by applying this inclusion–exclusion principle to the normal cycles of the simplices of the triangulation. Fu [46, 47, 50] showed that normal cycles could be defined for a very broad class of subsets called *geometric subsets*. In particular, semialgebraic sets, subanalytic sets, and more generally definable sets are geometric (see [12, 14, 15, 49] for the last point).

The main results of this chapter are two theorems on convergence and approximation for the normal cycles of sequences of triangulated polyhedra. The convergence theorem is a consequence of the compactness theorem for integral currents, under the assumption that the *fatness* of the triangulations is bounded from below [48]. The approximation theorem is a consequence of a deformation theorem of currents. Under a certain condition, we bound the difference of the curvature measures of two geometric sets when one of them is a smooth hypersurface. This result refines the theorem of Fu [48] by giving a quantitative version of it. More precisely, it gives an estimate of the flat norm of the difference between the normal cycle of a compact n-manifold K of $\mathbb{E}^n$ whose boundary is a smooth hypersurface and the normal cycle of a compact geometric subset $\mathcal{K}$, in terms of the mass of the normal cycle of K, the Hausdorff distance between their boundaries, the deviation angle between K and $\mathcal{K}$, and an a priori upper bound on the norm of the second fundamental form of the boundary of K.

Using these invariant forms and the normal cycle, Chap. 21 defines curvature measures of geometric sets, by integrating these forms on the normal cycles. Applying the convergence and approximation theorems of normal cycles, one deduces (by weak duality) convergence and approximation results of curvature measures of geometric sets [48]. This quantitative estimate of the difference between curvature measures of two "close" subsets generalizes those given for convex subsets.

Chapter 22 notes that the previous theory deals with principal curvatures but never with principal directions. To get a finer description of the geometry of singular sets, it is natural to look for a generalization of the second fundamental form of an immersion to the singular case. Mimicking the construction of the invariant $(N-1)$-forms, we define a $(0,2)$-tensor valued $(N-1)$-form that we plug in the normal cycle of the considered geometric subset $\mathcal{K}$. In this way, we create a new curvature measure which we call *the second fundamental measure associated to $\mathcal{K}$*. Of course, when $\mathcal{K}$ is smooth, we get the integral of the second fundamental form. As before, we deduce convergence and approximation theorems in terms of this new second fundamental measure.

1.10 Part VIII: Applications to Curves and Surfaces

Chapters 23–26 apply the results of the previous chapters to the most useful situations: curves and surfaces in $\mathbb{E}^2$ and $\mathbb{E}^3$. We give explicit computations and, when it is possible, explicit bounds on the approximations.

The last chapter (Chap. 27) is devoted to the applications of the previous theories to the Voronoi diagram and Delaunay triangulations. After a brief summary of the main constructions, in particular the construction of a restricted Delaunay triangulation associated to a curve of a surface, we deal with the approximation of the length, area, and curvatures of a sampled curve or surface.

To end this introduction, we would like to point out the fundamental difference between a *convergence* result and an *approximation* one: when one deals with applications (like medical imaging, structural geology, or computer graphics for instance), a *convergence* result of geometric invariants is often elegant and reassuring. But how to apply it? Conversely, an *approximation* result gives a bound on the error. However, in both cases, we are often dealing with a "real-world object," extremely difficult to define. We must have permanently in mind the difference between a "real" physical object, the perception of this object, and its mathematical modeling.

As an example, one of the plates presented in this book is a reconstitution of the principal directions of the head of Michelangelo's David. The validity of this image is implicitly admitted by the fact that one recognizes the Michelangelo masterpiece. But a basic problem is occulted: is it well founded to assign directions or lines of curvatures to an eventually smooth David, and then trying to approximate them by those of a triangulation sufficiently close to this hypothetical smooth surface?

Acknowledgments There are several people I would like to thank: E. Boix, V. Borrelli, B. Thibert, D. Cohen-Steiner (with whom I had long discussions), J. Fu (who introduced me to the subject), K. Polthier, J.D. Boissonnat, and the members of the Projet Geometrica (I.N.R.I.A.), N. Ayache (who encouraged me to write this book) and the members of the Projet Asclepios (I.N.R.I.A.), F. Chazal, T.K. Dey, P. Orro, and C. Grand. I also thank the language editor Prof. Michael Eastham (Cardiff University) who corrected my English grammar. Finally, I thank the referees who pointed out misprints and more serious mistakes in a previous version of the text.

Il y a entre les géomètres et les astronomes une sorte de malentendu au sujet de la signification du mot convergence. *Les géomètres, préoccupés de la parfaite rigueur et souvent trop indifférents à la longueur des calculs inextricables dont ils conçoivent la possibilité, sans songer à les entreprendre effectivement, disent qu'une série est convergente quand la somme des termes tend vers une limite déterminée, quand même les premiers termes diminueraient très lentement. Les astronomes, au contraire, ont coutume de dire qu'une série converge quand les vingt premiers termes, par exemple, diminuent très rapidement, quand même les termes suivants devraient croître indéfiniment. Ainsi pour prendre un exemple simple, considérons les deux séries qui ont pour terme général $\frac{1000^n}{n!}$ et $\frac{n!}{1000^n}$. Les géomètres diront que la première série converge, et même qu'elle converge rapidement,...;*

mais ils regarderont la seconde comme divergente. Les astronomes, au contraire, regarderont la première comme divergente,..., et la seconde comme convergente. Les deux règles sont légitimes: la première dans les recherches théoriques; la seconde dans les applications numériques...

Henri Poincaré,

(Méthodes nouvelles de la mécanique céleste, Chapitre 8 tome 2, 1884).

Part I
Motivations

Chapter 2
Motivation: Curves

The length and the curvature of a smooth space curve, the area of a smooth surface and its Gauss and mean curvatures, and the volume and the intrinsic (resp., extrinsic) curvatures of a Riemannian submanifold are classical geometric invariants. If one knows a parametrization of the curve (resp., the surface, resp., the submanifold), these geometric invariants can be directly evaluated. If such parametrizations are not given, one may approximate these invariants by approaching the curve (resp., the surface, resp., the submanifold), by suitable discrete objects, on which simple evaluations of these invariants can be done. Our goal is to investigate a framework in which a geometric theory of both smooth and discrete objects is simultaneously possible. To motivate this work, we begin with two simple examples: the length and curvature of a curve.

2.1 The Length of a Curve

This book deals essentially with curves, surfaces, and submanifolds of the Euclidean space $\mathbb{E}^N$ endowed with its classical scalar product $< .,. >$.

2.1.1 The Length of a Segment and a Polygon

If p and q are two points of $\mathbb{E}^N$, the *length* of the segment pq is the norm of the vector $\vec{pq}$, i.e., the real number

$$|pq| = \sqrt{< \vec{pq}, \vec{pq} >}.$$

If P is a polygon, given by a (finite ordered) sequence of points $v_1, ..., v_n$ in $\mathbb{E}^N$, the *length* $l(P)$ of P is the sum of the lengths of its edges, i.e.,

$$l(P) = \sum_{i=1}^{n-1} |v_i v_{i+1}|.$$

2.1.2 The General Definition

Let us now give the classical definition of the length of a curve, using approximations to the curve by polygons. The length of a curve (without any assumption on
regularity) is usually defined as the supremum of the lengths of all polygons inscribed in it: let

$$c : I = [a,b] \to \mathbb{E}^N$$

be a (parametrized) curve from a segment $[a,b] \subset \mathbb{R}$ into $\mathbb{E}^N$. If there is no possible
confusion, we identify the image Γ of c (i.e., the *support* of the curve c) with c
(Fig. 2.1).

Definition 1. Let $\mathcal{S}$ be the set of all finite subdivisions $\sigma = (t_0, t_1, ..., t_i, ..., t_n)$ of
$[a,b]$, with

$$a = t_0 < t_1 < ... < t_i < ... < t_n = b,$$

and denote by $l(\sigma)$ the length of the polygon $c(t_0)c(t_1)...c(t_i)...c(t_n)$. If

$$\sup_{\sigma \in \mathcal{S}} l(\sigma)$$

is finite, one says that the curve c is rectifiable and its length $l(c)$ (or $l(\Gamma)$) is this
supremum:

$$l(c) = \sup_{\sigma \in \mathcal{S}} l(\sigma). \qquad (2.1)$$

It is well known that there exist continuous curves which are not rectifiable. The
most famous example is the *Von Koch curve* obtained as follows: start from an equilateral triangle and consider each of its edges e. Take off the middle third e_1 of
e and replace it with an equilateral triangle t_1. Then, take off e_1. The limit of this
process gives rise to the *Von Koch curve*, which is continuous but with infinite length
(Fig. 2.2).

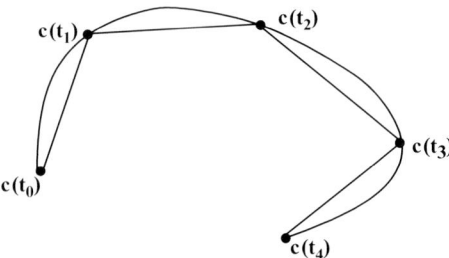

Fig. 2.1 A smooth curve and its
approximation by a polygon line

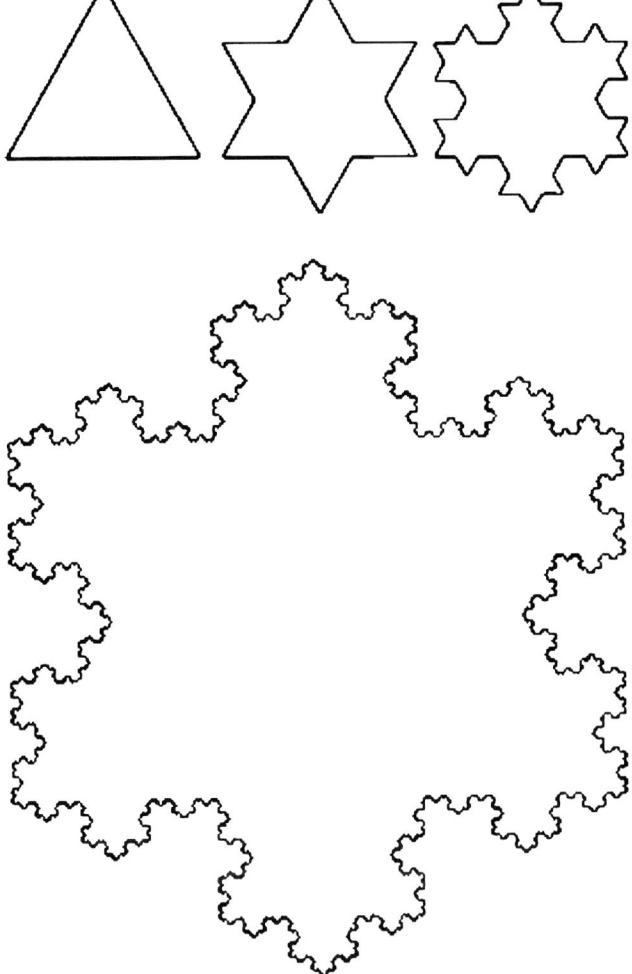

Fig. 2.2 Von Koch curve is continuous but not rectifiable

2.1.3 The Length of a C^1-Curve

On the other side of regularity, a classical theorem asserts that C^1-curves are rectifiable (see [78] for instance). This theorem is a consequence of the mean value theorem. This strong assumption implies an expression for the length in terms of the integral of the norm of its speed vector field:

$$l(c) = \int_a^b |c'(u)| du, \tag{2.2}$$

where $\int$ means the Riemann integral.

2.1.4 An Obvious Convergence Result

By definition, one "approaches" the length of a smooth curve by inscribing a polygon on the curve and evaluating the length of the polygon. The following result is a simple consequence of the definition.

Theorem 1. *Let* $\sigma^k = (t_0^k, t_1^k, \ldots, t_i^k, \ldots, t_{n_k}^k)_{k \in \mathbb{N}}$ *be a sequence of subdivisions of a segment* $[a,b]$, *with*

$$a = t_0^k < t_1^k < \ldots < t_i^k < \ldots < t_n^k = b.$$

Let

$$c : [a,b] \to \mathbb{E}^N$$

be a rectifiable curve and denote by $l(\sigma^k)$ *the length of the polygon* P_k *of* $\mathbb{E}^N$ *defined by* $c(t_0^k)c(t_1^k)\ldots c(t_i^k)\ldots c(t_{n_k}^k)$. *If the length of the edges of* P_k *tends to 0 when k tends to* $+\infty$, *then*

$$\lim_{k \to \infty} l(P_k) = l(c).$$

2.1.5 Warning! Negative Results

- Note that Theorem 1 needs to ensure that the vertices of the polygons are *on* the curve. If one only assumes that they are "close" to the curve, the result fails. Figure 2.3 shows a sequence of polygons of length $4\sqrt{2}$ tending (for the Hausdorff topology) to a straight line of length 4 (the polygon lines are not *inscribed* on the straight line).
- On the other hand, note that this convergence result is true because we have assumed an order on the vertices of the polygons. It is clear that if we change this order, creating new edges and canceling others, the length of the resulting sequence of polygons does not converge in general to the length of the curve (see Fig. 2.4).

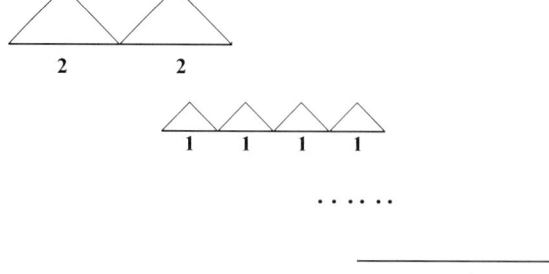

Fig. 2.3 In general, the length functional is not continuous

Fig. 2.4 In this example, the sequence $t_0, t_1, \ldots, t_4$ is not increasing and the sequence of lengths of such polygons may not converge to the length of the curve

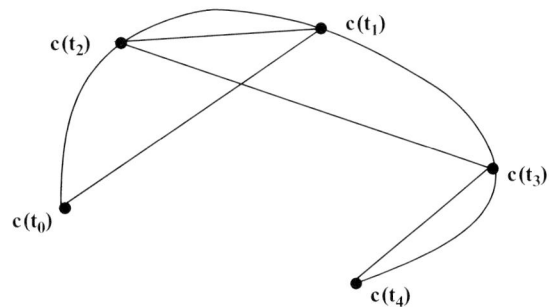

2.2 The Curvature of a Curve

Although one usually defines the length of a smooth curve as a limit of the length of polygons inscribed in it, one defines the curvature of a smooth curve differentiating its tangent vector field. We recall here the classical definition of the curvature of a smooth curve and the corresponding definition for polygons.

2.2.1 The Pointwise Curvature of a Curve

1. *The pointwise curvature of a C^2-curve.* Consider a C^2 *regular* curve

$$c : I \to \mathbb{E}^N$$

(for every u in the interval I, $c'(u) \neq 0$). We know that c admits a parametrization

$$\gamma : [0, l] \to \mathbb{E}^N,$$

by the arc length s, i.e., $|\gamma'(s)| = 1$, where l denotes the length of the curve. Let t denote its (unit) tangent vector field, i.e., $t = \gamma'$. At a point $m = \gamma(s)$, the *curvature $k(m)$ of the curve*[1] is the norm of the derivative t' of t (t' is orthogonal to t since $<t, t> = 1$):

$$k(m) = |t'(s)|. \tag{2.3}$$

This definition implies that the curvature is a nonnegative function defined on the curve (Fig. 2.5).
Another equivalent definition of the curvature of a curve γ at $m = \gamma(s)$ is given by

$$k(m) = \lim_{h \to 0^+, k \to 0^+} \frac{\angle(\gamma'(s-h), \gamma'(s+k))}{h+k}, \tag{2.4}$$

---

[1] If no confusion is possible, we can write $k(s)$ instead of $k(m)$ when $m = \gamma(s)$.

Fig. 2.5 The orthogonal
frame (t,t') over a point of a
plane curve

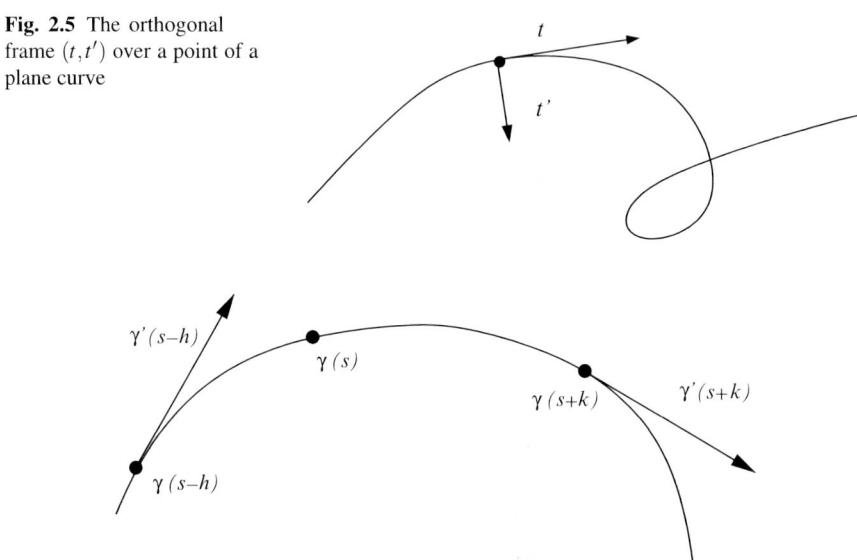

Fig. 2.6 Pointwise convergence

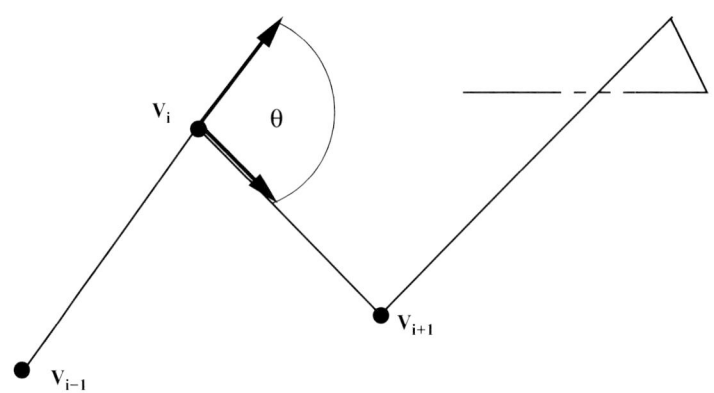

Fig. 2.7 The angle between two incident edges

where $\angle(\gamma'(s-h),\gamma'(s+k))$ denotes the angle ($\in [0,\pi]$) between the tangent
vectors of the curve at the points $\gamma(s-h)$ and $\gamma(s+k)$ (see [41] for instance).
We shall make precise this point of view for *plane* curves in Sect. 2.4 (Fig. 2.6).

2. *The curvature of a polygon at one of its vertices.* By analogy with (2.4), one can
   define the curvature of a polygon in $\mathbb{E}^N$ at one of its vertices as the angle of its
   incident edges (Fig. 2.7).

Definition 2. Let $P = v_1 v_2 ... v_n$ be a polygon in $\mathbb{E}^N$. The curvature of P at each
interior vertex v_i is the real number

$$k_{v_i} = \angle(\overrightarrow{v_{i-1}v_i}, \overrightarrow{v_iv_{i+1}}).$$

For further use, note that each angle is also the angle between the (correctly oriented) normals of the corresponding edges.

To end this section, we note that the pointwise curvature of a smooth curve has the dimension of the inverse of a length. On the other hand, the pointwise curvature of a polygon is dimensionless, since it is an angle.[2] However, we shall see in Sect. 2.2.2 that the global (or total) curvature of a smooth curve and a polygon are both dimensionless.

2.2.2 The Global (or Total) Curvature

1. *The case of a smooth curve.* Let

$$\gamma : [0,l] \rightarrow \mathbb{E}^N$$

be a smooth regular curve. With the usual notation, one defines the global curvature by integration as follows.

Definition 3. The global curvature of a smooth curve γ is the real number

$$\mathcal{K}(\gamma) = \int_0^l k(s)ds.$$

The global curvature of a curve is also called the *total curvature*.

2. *The case of a polygon.* By analogy with (2.4), one can define the *global (or total) curvature* of a polygon in $\mathbb{E}^N$ as the sum of the angles of its consecutive edges. With the usual notation:

Definition 4.
$$\mathcal{K}(P) = \sum_{i=1}^{i=n-2} \angle(\overrightarrow{v_iv_{i+1}}, \overrightarrow{v_{i+1}v_{i+2}}),$$

where each angle belongs to $[0,\pi]$.

Using (2.4), Fary [41] was probably the first to prove that the total curvature of a curve is the limit of the sum of the angles of a sequence of inscribed polygons: using the theory of the Stieltjes integral, one deduces from (2.4) that

$$\int_0^l k(s)ds = \lim_{|s_i - s_{i+1}| \to 0} \sum_{s_i} \angle(\gamma'(s_i), \gamma'(s_{i+1})). \qquad (2.5)$$

This can be interpreted as follows.

---

[2] That is why some authors prefer to define the pointwise curvature at a vertex of a polygon by dividing the angle by a length, like half the sum of the lengths of the incident edges for instance.

Theorem 2. *Let* $\sigma^k = (t_0^k, t_1^k, ..., t_i^k, ..., t_{n_k}^k)_{k \in \mathbb{N}}$ *be a sequence of subdivisions of a segment* $[a, b]$, *with*

$$a = t_0^k < t_1^k < ... < t_i^k < ... < t_n^k = b.$$

Let

$$\gamma : [a, b] \to \mathbb{E}^N$$

be a smooth regular curve of $\mathbb{E}^N$ *(parametrized by the arc length). Let* $(P_k)_{k \in \mathbb{N}}$ *be the sequence of polygons inscribed in the curve* γ *of* $\mathbb{E}^N$ *defined by* $c(t_0^k)c(t_1^k)...c(t_i^k)...$ $c(t_{n_k}^k)$. *If the length of the edges of* P_k *tends to 0 when k tends to infinity, then the global curvature of* P_k *tends to the global curvature of* γ:

$$\lim_{k \to \infty} \mathcal{K}(P_k) = \mathcal{K}(\gamma).$$

In some sense, this means that the sum of the angles between consecutive edges of a polygon inscribed in the curve gives an approximation of its global curvature.

Let us now mention the famous theorem of Fenchel and Milnor, concerning smooth or polygonal curves [44]. In our context, this theorem can be stated as follows.

Theorem 3. *Let C be a smooth or polygonal closed curve in* $\mathbb{E}^N$. *Then, the total curvature of C is larger than or equal to* 2π:

$$\mathcal{K}(C) \geq 2\pi, \tag{2.6}$$

with equality if and only if the curve is planar and convex.

This theorem has been improved for knotted curves by Fary [41] and Milnor [60]: if the curve is knotted, then

$$\mathcal{K}(C) \geq 4\pi. \tag{2.7}$$

It is interesting to note that Milnor's proofs (of Theorem 3) use both a discrete and a smooth approach, approximating smooth curves by polygons *inscribed in them* (Fig. 2.8).

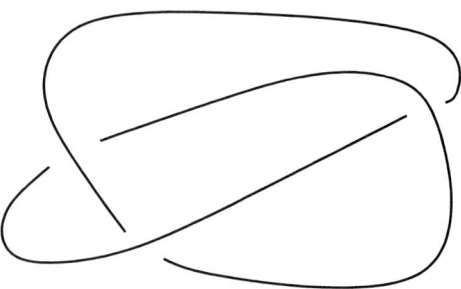

Fig. 2.8 The total curvature
of this (knotted) curve is $\geq 4\pi$

2.3 The Gauss Map of a Curve

Let us now introduce the Gauss map, with which we can recover the global curvature. Let

$$\gamma : [0,l] \to \mathbb{E}^N$$

be a smooth curve.

Definition 5. The map

$$\begin{aligned} \mathbf{G} : [0,l] &\to \mathbb{S}^{N-1}, \\ s &\to \gamma'(s) \end{aligned}$$

is called the Gauss map associated to γ.

Since $|\mathbf{G}'| = k$, the length of the curve $\mathbf{G}([0,l])$ is nothing but the global curvature $\mathcal{K}(\gamma)$ of γ (Fig. 2.9).

This construction can be generalized to polygons as follows: if P is a polygon in $\mathbb{E}^N$, one can define its Gauss map $\mathbf{G}$ as the map which associates to each edge e of P the unit vector of $\mathbb{S}^{N-1}$ parallel to e (with the correct orientation). The image of $\mathbf{G}$ is nothing but a finite set of points which can be joined by arcs of great circles. One gets a curve on $\mathbb{S}^{N-1}$ whose length is nothing but the global curvature of P, as defined in Definition 4 (Fig. 2.10).

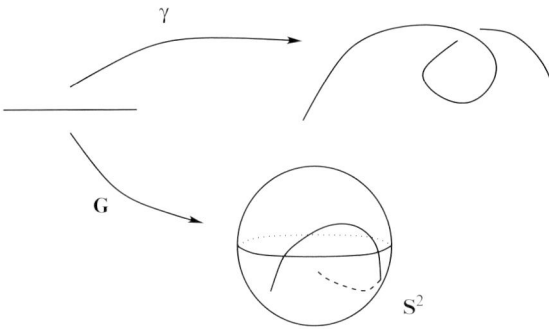

Fig. 2.9 The Gauss map associated to a curve in $\mathbb{E}^3$

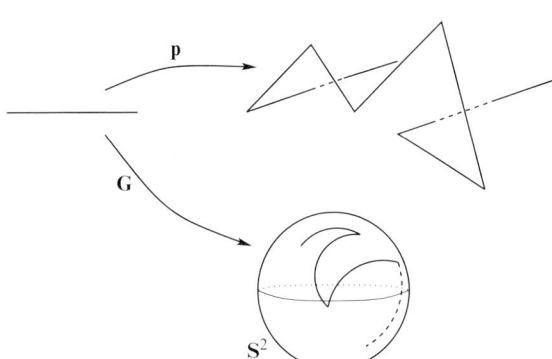

Fig. 2.10 The Gauss map associated to a polygon line

Consequently, it appears that one can *approach* the global curvature of a curve by evaluating the length of the image of the Gauss map of a polygon inscribed in it. Detailed computations will be made in Chap. 25 for curves in $\mathbb{E}^2$ and $\mathbb{E}^3$. One might think that this method can be applied for any geometric invariant defined on any submanifold of $\mathbb{E}^N, N \geq 3$. In fact, the situation is completely different even for surfaces in $\mathbb{E}^3$, as we shall show in the next chapters.

2.4 Curves in $\mathbb{E}^2$

One can get sharper results and definitions for plane curves, as we see in this section.

2.4.1 A Pointwise Convergence Result for Plane Curves

The curvature k_p at a point p of a C^2-plane curve γ can be evaluated from the angular defect of a polygon inscribed in it as follows.

Theorem 4. *Let $p_1 p p_2$ be three points on γ, and denote by η_1 (resp., η_2) the distance from p_1 to p (resp., p_2 to p) and by α the angle between pp_1 and pp_2. Then:*

- *If $\eta_1 \neq \eta_2$,*

$$\frac{\pi - \alpha}{\overline{\eta}} = k_p + o(1).$$

- *If $\eta_1 = \eta_2 = \eta$,*

$$\frac{\pi - \alpha}{\eta} = k_p + o(\eta),$$

  where

$$\overline{\eta} = \frac{\eta_1 + \eta_2}{2}.$$

This theorem can be proven by using a power expansion of the parametrization of the curve (see [21] for details). Note in particular that the convergence speed is faster when the two neighbors are located at the same distance from p (Fig. 2.11).

2.4.2 Warning! A Negative Result on the Approximation by Conics

Theorem 4 claims that one can approximate the pointwise curvature of a curve by the angle spanned by two inscribed segments. It may be tempting to get a better approximation by using an inscribed conic instead of segments. If one wishes to approximate the curvature of a smooth curve at a point p, one could adopt the following process:

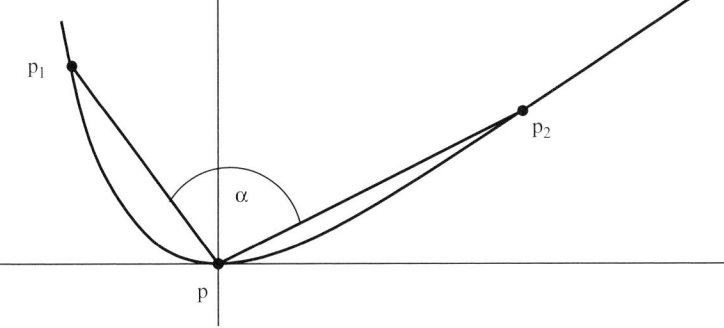

Fig. 2.11 The pointwise curvature of a smooth curve approximated by the angle

- Consider four points q, r, s, t close to p.
- Construct the conic $\mathcal{C}$ on p, q, r, s, t.
- Compute the exact value of the curvature κ_p of $\mathcal{C}$ at p.
- Approximate the curvature k_p of γ at p by κ_p.

These steps are not valid in general, as the following example[3] shows.

A counterexample. Consider the function

$$f : \mathbb{R} \to \mathbb{R},$$

defined by

$$f(x) = \begin{cases} g(x)\sqrt{x^2 + x} & \text{if } x \geq 0, \\ 0 & \text{if } x < 0, \end{cases} \tag{2.8}$$

where

$$g : \mathbb{R} \to \mathbb{R}$$

is a smooth function such that

$$g(x) = \frac{1}{2^n}$$

on the segment $\left[\frac{1}{2n+1}, \frac{1}{2n}\right]$.

One can check that the function g can be chosen so that f is C^∞ (the delicate point is the origin 0), has a horizontal tangent at 0, and zero curvature at 0. To construct an explicit suitable function g, one begins considering a C^∞ map h defined on the segment $[0, 1]$, such that $h(0) = 0$ and $h(1) = 1$, whose derivatives are all null at 0 and 1. Then, we put

$$g(x) = \begin{cases} 0 & \text{if } x = 0; \\ \frac{1}{2^{n+1}} + \frac{1}{2^{n+1}}h((2n+1)(2n+2)x - (2n+1)), & \text{on } \left[\frac{1}{2n+2}, \frac{1}{2n+1}\right]; \\ \frac{1}{2^n} & \text{on } \left[\frac{1}{2n+1}, \frac{1}{2n}\right]. \end{cases} \tag{2.9}$$

---

[3] Due to Pierre Lavaurs in a private communication.

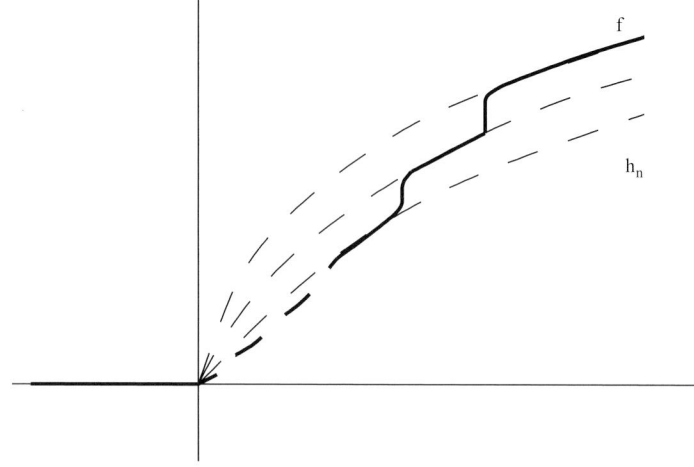

Fig. 2.12 At the origin, the curvature of the sequence of dotted hyperbolae tends to infinity, although the curvature of the plain curve is null

The only difficulty is to check that g is C^∞ at 0 and that all its derivatives are zero at this point.

Now for each $n \in \mathbb{N}$, consider four points a_n, b_n, c_n, d_n on the graph of f between $\frac{1}{2n+1}$ and $\frac{1}{2n}$. Since g equals $\frac{1}{2^n}$ on this segment, the five points $0, a_n, b_n, c_n, d_n$ lie on the hyperbola

$$y = \frac{1}{2^n} \sqrt{x^2 + x},$$

whose tangent at 0 is vertical and whose curvature at 0 equals $\frac{4^n}{2}$.

When n goes to infinity, we get a sequence of four points a_n, b_n, c_n, d_n which lie on the graph of f, which tends to 0, and such that the associated conic through $0, a_n, b_n, c_n, d_n$ has curvature at 0 tending to infinity, although the curvature of the graph of f at 0 is null (Fig. 2.12).

Other types of counterexamples can be constructed by the reader, replacing the hyperbola by ellipses for instance. The curvature k_p at the relevant point p on the original curve can be zero or not, and the curvature of the sequence of ellipses tend to a (zero or nonzero) value different to k_p.

2.4.3 The **Signed** *Curvature of a Smooth Plane Curve*

If the smooth curve γ lies in $\mathbb{E}^2$ endowed with its canonical orientation, one can modify slightly the definition of k to define the *signed* curvature κ at each point $m = \gamma(s)$ as follows: for $s \in [0, l]$, one defines the (unit) normal vector $v(s)$ such that $(t(s), v(s))$ is a direct orthonormal frame at s. We put (Fig. 2.13)

$$\kappa(m) = \langle t'(s), v(s) \rangle. \tag{2.10}$$

Fig. 2.13 There are two possible definitions of the curvature of a plane curve but the functions k and κ may only differ in sign

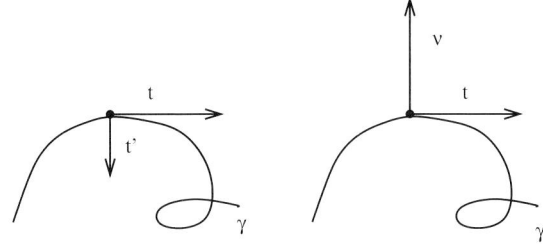

Fig. 2.14 Locally, θ measures the angle between the x-axis and the tangent t of the curve

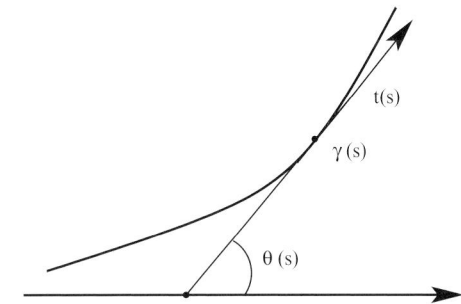

Then, as before, we define the global (or total) curvature $\tilde{\mathcal{K}}(\gamma)$ of γ by

$$\tilde{\mathcal{K}}(\gamma) = \int_0^l \kappa(s)ds.$$

One can give an intuitive geometric meaning of the (signed) curvature of a curve: let

$$\gamma : [0,l] \longrightarrow \mathbb{E}^2$$
$$s \longrightarrow \gamma(s)$$

be a regular curve parametrized by the arc length and t be its (unit) tangent vector field. If ε is a sufficiently small real number, let us define the map

$$\theta : [0,\varepsilon] \longrightarrow [0,2\pi[$$
$$s \longrightarrow \theta(s),$$

which associates to each $s \in [0,\varepsilon]$ the angle $\theta(s) \in [0,2\pi[$ which $t(s)$ makes with the Ox-axis in $\mathbb{E}^2$ (Fig. 2.14).
Then

$$t = (\cos\theta, \sin\theta).$$

If v is a vector field over γ such that (t,v) is a direct orthonormal frame, then

$$t' = \theta'(s)v = \kappa v,$$

where κ denotes the *signed* curvature function of the curve. We deduce that, for each $s_0 \in [0,\varepsilon]$,

$$\theta(s_0) = \int_0^{s_0} \kappa(s)ds. \tag{2.11}$$

Consequently, the integral of κ on $[0, \varepsilon]$ is the variation of the angle θ:

$$\int_0^\varepsilon \kappa(s)ds = \theta_\varepsilon - \theta_0.$$

This implies that the integral of κ is perfectly known if one knows the position of the tangent vectors at the end points of γ.

Globally, let us introduce the *Gauss map* **G** associated to γ:

$$\mathbf{G} : [0,l] \to \mathbb{S}^1,$$

defined by

$$\mathbf{G}(s) = t(s), \forall s \in [0,l],$$

where t denotes the unit tangent vector field over the curve. If $d\theta$ denotes the length form of $\mathbb{S}^1$, one has

$$\mathbf{G}^*(d\theta) = \kappa ds, \qquad\qquad (2.12)$$

from which we deduce that

$$\tilde{\mathcal{K}}(\gamma) = \int_0^l \kappa ds = 2\mathbf{I}\pi, \qquad\qquad (2.13)$$

where κ denotes the *signed* curvature of the curve γ and **I** is an integer called the *rotation number* of γ. The *rotation number* is the number of times that the tangent vector field turns around $\mathbb{S}^1$ (the sign depending on the orientation of the curve).

Let us end this section with the well-known theorem.

Theorem 5 of Turning Tangents *The rotation number of a smooth closed simple*[4] *plane curve equals ± 1 (the sign depending on the orientation).*

2.4.4 The Signed *Curvature of a Plane Polygon*

One can also assign to each vertex of an (oriented) polygon P of the (oriented) plane $\mathbb{E}^2$ a *signed* angle as follows: consider any fixed axis, for instance the Ox-axis in $\mathbb{E}^2$. If the interior vertex v of P is incident to the (oriented) edges $\overrightarrow{pq}$ and $\overrightarrow{qr}$, let

$$\theta(\overrightarrow{pq}) \in]-\pi, \pi[$$

be the angle $\angle(Ox, \overrightarrow{pq})$ and

$$\theta(\overrightarrow{qr}) \in]-\pi, \pi[$$

be the angle $\angle(Ox, \overrightarrow{qr})$. We put

$$\theta_v = \theta(\overrightarrow{pq}) - \theta(\overrightarrow{qr}).$$

---

[4] A closed simple curve of the plane is a closed curve without self-intersection points.

Definition 6. Let $P = v_1v_2...v_n$ be an (oriented) polygon in the (oriented) plane $\mathbb{E}^2$. The *signed* curvature of P at each interior vertex v_i is the real number

$$\kappa_{v_i} = \theta_{v_i}.$$

As before, we define the global (or total) *signed* curvature of P by

$$\tilde{\mathcal{K}}(P) = \sum_i \kappa_{v_i}.$$

Since the sum of the signed curvatures of the vertices of a closed polygon $P = v_1...v_i...v_n$ of $\mathbb{E}^2$ is a multiple of 2π, the *rotation number* of P is the integer **I** defined by the equation

$$\tilde{\mathcal{K}}(P) = \sum_i \kappa_{v_i} = 2\mathbf{I}\pi.$$

The theorem of turning tangents for simple closed polygons of the plane (corresponding to Theorem 5) can be stated as follows.

Theorem 6. *The rotation number of a simple closed polygon equals* ±1 *(the sign depending on the orientation).*

The reader can consult [11, 40] for details.

2.4.5 Signed Curvature and Topology

It is important to remark that the *rotation number* **I** of a closed plane curve is invariant under isotopy. Consequently, the integral of the *signed* curvature κ of smooth or polygonal closed plane curves is a topological invariant. On the other hand, the *unsigned* curvature k is not a topological invariant: an isotopy of a closed plane curve changes its total curvature (Fig. 2.15).

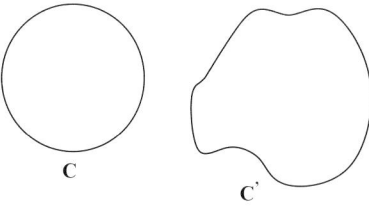

Fig. 2.15 An isotopy of the circle C does not change its rotation number (here equal to 1) nor its total *signed curvature*. However, its total curvature changes: the total *signed curvature* of C equals 2π and equals the total *signed curvature* of C'; the total curvature of C is still 2π but it is different to the total curvature of C'

2.5 Conclusion

In this chapter of motivation, we have seen that the *length* and the *curvature* invariants of curves can be defined in smooth and discrete contexts. However, even in simple situations, convergence and approximation results may fail.

The *length* and the *curvature* are the only two geometric invariants of curves we study in this book. They involve the first and second derivatives of the parametrization. Of course, other curvatures can be defined in a smooth context. For instance, let us write the *Frénet equations* for a (regular) smooth curve γ in $\mathbb{E}^3$ parametrized by the arc length, with (unit) tangent vector t:

$$\begin{cases} \frac{dt}{ds} & = kn \\ \frac{dn}{ds} & = -kt + \tau b \\ \frac{db}{ds} & = -\tau n, \end{cases} \tag{2.14}$$

from which one builds the Frénet frame (t, n, b). Here, τ denotes the torsion of the curve and b the binormal vector field. The torsion τ involves the *third* derivative of the parametrization. We do not deal with this invariant, although a nice theory should be presented.

Chapter 3 will deal with surfaces, where it appears that the theory of approximation of geometric invariants is much more complex.

Chapter 3
Motivation: Surfaces

Our goal in this chapter is to point out the difficulties arising when one evaluates the area and the curvatures of a surface by approximation.

3.1 The Area of a Surface

Let us begin with the area of a surface. We only assume here that the reader is familiar with the usual notion of area of simple linear objects like triangles and C^1-parametrized surfaces. A deeper summary of measure theory will be given in Chap. 5 (Fig. 3.1).

3.1.1 The Area of a Piecewise Linear Surface

To compute the area of a piecewise linear two-dimensional region, one divides the region into a partition of triangles, computes the area of each triangle with the familiar formula – *half the product of the base by the height* – and then adds all these areas. The only point to check is that the result is independent of the triangulation (Fig. 3.2).

3.1.2 The Area of a Smooth Surface

If a (bounded) surface S is the image of a parametrization (smooth almost everywhere),

$$x : U \to \mathbb{E}^3,$$

Fig. 3.1 The area of the triangle is $\frac{bh}{2}$

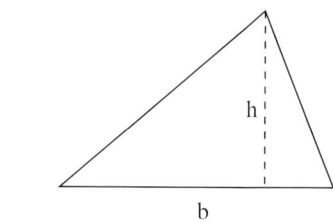

Fig. 3.2 The area of T is the sum of the areas of $t_1, t_2,$ and t_3

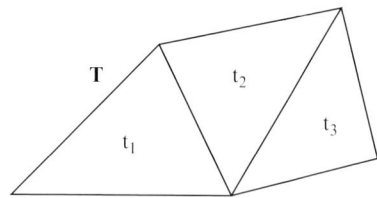

then the area of S is given by the integral formula

$$A(S) = \int_S \sqrt{dxdx}dudv, \tag{3.1}$$

denoting by (u,v) the coordinates of U. If S is piecewise linear, (3.1) obviously gives the exact result.[1]

3.1.3 Warning! The Lantern of Schwarz

From Sects. 3.1.1 and 3.1.2, one could think that the area of a smooth surface can be computed as the limit of the areas of a sequence of triangulations "tending to it." This section deals with the *noncontinuity* of the area with respect to the Hausdorff distance: one can construct examples of sequences of polyhedra P_n *inscribed* in a smooth surface S in E^3, whose Hausdorff limit is S, but whose areas do not tend to the area of S (compare with the case of a smooth curve approximated by a sequence of *inscribed* polygons described in Chap. 2). The classical example of such a situation is the *Lantern of Schwarz* [75], which we describe now in details (see also [11]).

Let C be a cylinder of finite height l and radius r in $\mathbb{E}^3$. Let $P(n,m)$ be the triangulation inscribed in C defined as follows: consider $m+1$ circles on the cylinder C obtained by intersecting C with 2-planes orthogonal to the axis of C. Inscribe in each circle a regular n-gon such that the n-gon on the slice k is obtained from the n-gon of the slice $k-1$ by a rotation of angle $\frac{\pi}{n}$. Then, join each vertex v of the slice $k-1$ to the two vertices of the slice k which are nearest to v. One obtains a

---

[1] Denoting $\sqrt{dxdx}$ by $|dxdx|$, (3.1) can be written $A(S) = \int_S |dx|dudv$.

Fig. 3.3 Hexagons inscribed in a circle

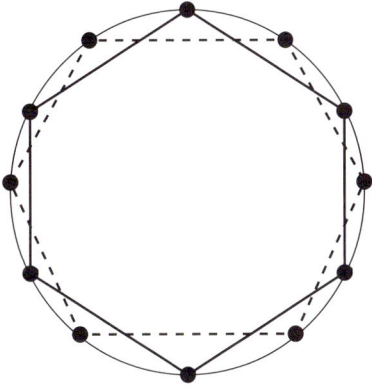

Fig. 3.4 The Lantern of Schwarz

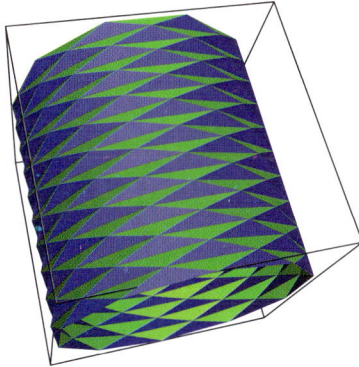

triangulation whose vertices v_{ij} are defined as follows: for all $i \in \{0,...,n-1\}$ and for all $j \in \{0,...,m\}$ (Figs. 3.3 and 3.4),

$$v_{ij} = \begin{cases} (r\cos(i\alpha), r\sin(i\alpha), jh) & \text{if } j \text{ is even,} \\ (r\cos(i\alpha + \frac{\alpha}{2}), r\sin(i\alpha + \frac{\alpha}{2}), jh) & \text{if } j \text{ is } odd, \end{cases}$$

and whose faces are

$$v_{ij}\, v_{(i+1)j}\, v_{i(j+1)},$$
$$v_{ij}\, v_{(i-1)(j+1)}\, v_{i(j+1)},$$

where $\alpha = \frac{2\pi}{n}$ and $h = \frac{l}{m}$.

Note now that the area of the cylinder is $\mathcal{A}(C) = 2\pi rl$. On the other hand, the area $\mathcal{A}(P(n,m))$ of $P(n,m)$ is nothing but the sum of the areas of its triangles. When n tends to infinity, a simple computation shows that

$$\lim_{n\to+\infty} \mathcal{A}(P(n,m)) = 2\pi rl\sqrt{1 + \frac{m^2\pi^4 r^2}{4n^4 l^2}}. \tag{3.2}$$

In particular,

$$\lim_{n \to +\infty} \mathcal{A}(P(n,n^2)) = 2\pi r l \sqrt{1 + \frac{r^2 \pi^4}{4l^2}} \neq \mathcal{A}(C) \tag{3.3}$$

and

$$\lim_{n \to +\infty} \mathcal{A}(P(n,n^3)) = +\infty, \tag{3.4}$$

although C is the Hausdorff limit of both $P(n,n^2)$ and $P(n,n^3)$.

This example shows that it is possible to find sequences of triangulations inscribed in a (smooth) surface S whose Hausdorff limit is C, but whose area tends to infinity or to a limit different from the area of C. One can visualize this phenomenon by building developable *Lanterns of Schwarz* as Figs. 3.5 and 3.6 show.

More generally, it can be proved that for every real number $\alpha > 2\pi r l$, there exists a sequence of Schwarz lanterns whose Hausdorff limit is C and whose areas tend to α.

The consequence of these crucial remarks is that the Hausdorff topology is not the best one to deal with geometric approximations.

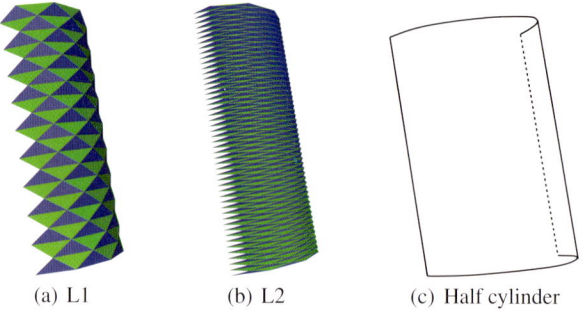

(a) L1   (b) L2   (c) Half cylinder

Fig. 3.5 Examples of *half Schwarz lanterns*

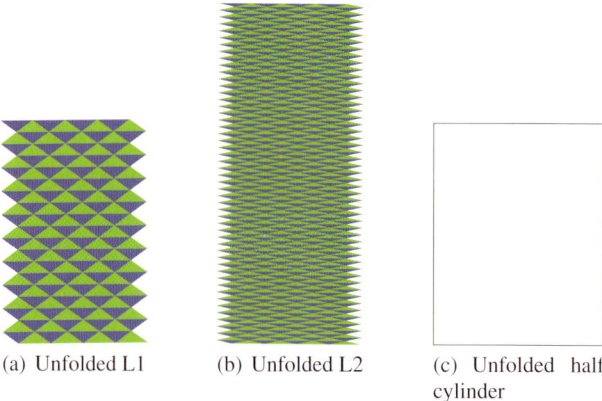

(a) Unfolded L1   (b) Unfolded L2   (c) Unfolded half cylinder

Fig. 3.6 Unfolding of C and of two *half Schwarz lanterns* closely inscribed in C (the scale is the same)

3.2 The Pointwise Gauss Curvature

After giving the basic definition of the Gauss curvature at a point of a smooth surface or a vertex of a polyhedron, this section focusses on the difficulty of getting
pointwise approximations or convergence theorems.

3.2.1 Background on the Curvatures of Surfaces

Let S be a (smooth) oriented surface of the (oriented) Euclidean space $\mathbb{E}^3$. The surface S can be endowed with the Riemannian metric induced by the canonical scalar
product of $\mathbb{E}^3$. Classically, the *Gauss curvature function* G of S can be defined with
the metric on S. In our context, we prefer to define it by using the *extrinsic* structure
of S. *Theorema egregium* of Gauss asserts that both definitions are equivalent.

We denote by TS the tangent bundle of S and by ξ the unit normal vector field of
S. If p is a point of S, S can be locally defined around m by a smooth immersion

$$x : U \to \mathbb{E}^3,$$

where U is a domain of $\mathbb{E}^2$. Let (u, v) be a system of coordinates on U.

Since $|\xi| = 1$,

$$\left(\frac{\partial \xi}{\partial u}\right)_p \text{ and } \left(\frac{\partial \xi}{\partial v}\right)_p$$

are orthogonal to ξ_p and define a frame of T_pS. Consequently, the differential $D\xi$
induces a tensor

$$A : TS \to TS,$$

defined for each vector $X \in TS$ by

$$A(X) = -D_X \xi.$$

The tensor A is self-adjoint with respect to the scalar product $< \ , \ >$: for every
$X, Y \in TS$,

$$< A(X), Y >=< X, A(Y) > .$$

The tensor A is called the *Weingarten tensor* and its adjoint h, defined for every
$X, Y \in TS$ by

$$h(X, Y) =< A(X), Y >,$$

is called the *second fundamental form* of S. At each point p of S, the eigenvalues
λ_{1_p} and λ_{2_p} of h at p are called the principal curvatures of S at p. At each point p,
there exist two (orthogonal) eigenvectors of A, called the *principal directions* of S
at p. The integral lines of these principal directions are called the *lines of curvature*
of S.

Fig. 3.7 At the point p, the two principal lines (tangent to the principal directions) are orthogonal

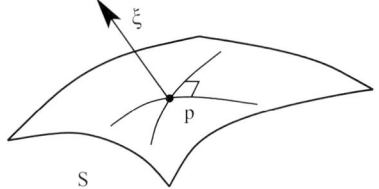

The determinant of h is the Gauss curvature G_p of S at p and half the trace H_p of h is the mean curvature of S at p:

$$G_p = \lambda_{1_p}\lambda_{2_p},$$
$$H_p = \tfrac{1}{2}(\lambda_{1_p} + \lambda_{2_p}).$$

Another (equivalent) point of view is to define the principal curvatures and directions as follows: a *geodesic* of S is a curve minimizing the distance locally. It is well known that, at each point p of S and for any (unit) tangent direction X at p, there exists a unique (local) geodesic γ_X tangent to X. When X varies on the unit circle $\mathbb{S}^1 \subset T_p S$, the curvature $k_p(X)$ at p of γ_X (considered as a curve in $\mathbb{E}^3$) is related to the second fundamental form as follows:

$$k_p(X) = h_p(X,X).$$

The maximum and minimum of $k_p(X)$ are the two principal curvatures λ_{1_p} and λ_{2_p} of S at p. An extensive study of the local Riemannian geometry of surfaces in $\mathbb{E}^3$ can be found in [40] (Fig. 3.7).

3.2.2 Gauss Curvature and Geodesic Triangles

Let us present a discrete point of view: it is possible to recover the Gauss curvature at a point p of a smooth surface S with help of the *solid angle* of *geodesic triangles* incident to p (see [2] for a extensive study). If v, v_1, v_2 are three points on S, the geodesic triangle with vertices v, v_1, v_2 is the union of the geodesic arc γ_1 joining v and v_1, the geodesic arc γ_2 joining v and v_2, and the geodesic arc γ_3 joining v_1 and v_2. If τ is a *geodesic triangle* on S, with vertices v, v_1, v_2, denote by l, l_1, l_2 the lengths of the geodesic arcs opposite to the vertices v, v_1, v_2. Finally, let α be the angle of τ at v.

From this geodesic triangle, we construct now an Euclidean triangle t whose vertices are p, p_1, p_2 in $\mathbb{E}^2$ and whose edge lengths are

$$l = |p_1 p_2|, l_1 = |p p_2|, l_2 = |p p_1|.$$

Let β be the angle of t at p. A fundamental result can be stated as follows.

Lemma 1. *The formula*

$$\beta = \alpha + \frac{1}{6}G_p l_1 l_2 \sin\alpha + o(\bar{l}^2) \tag{3.5}$$

holds, where $\bar{l} = \max(l_1, l_2).$[2]

Lemma (3.5) means that the Gauss curvature of a surface S at a point can be approximated (up to order 2) by (a constant times) a difference of two angles $\beta - \alpha$. Note, however, that α is *not* the angle at v of the triangle composed of the *chord* – i.e., straight lines – (in $\mathbb{E}^3$) joining v, v_1, v_2. The quantities α and β only depend on the *intrinsic* Riemannian geometry of S and not on the isometric embedding of S in $\mathbb{E}^3$.

Consider now a geodesic triangulation around p, i.e., a set of n geodesic triangles incident to p, forming a topological disc around p. Summing (3.5) over all angles β_i at p and using the fact that the sum of the β_i equals 2π, one gets [2]:

Theorem 7.

$$2\pi - \sum_i \alpha_i = \frac{\mathcal{A}(p)}{3}G_p + o(l^2), \tag{3.6}$$

where G_p *is the Gauss curvature of* S *at* p *and* $\mathcal{A}(p)$ *denotes the sum of the areas of the Euclidean triangles* t_i *associated to the geodesic triangles* τ_i *(Figs. 3.8 and 3.9).*

This theorem is the heart of [24, 25] in which a deep intrinsic discretization of the curvature tensor is given.

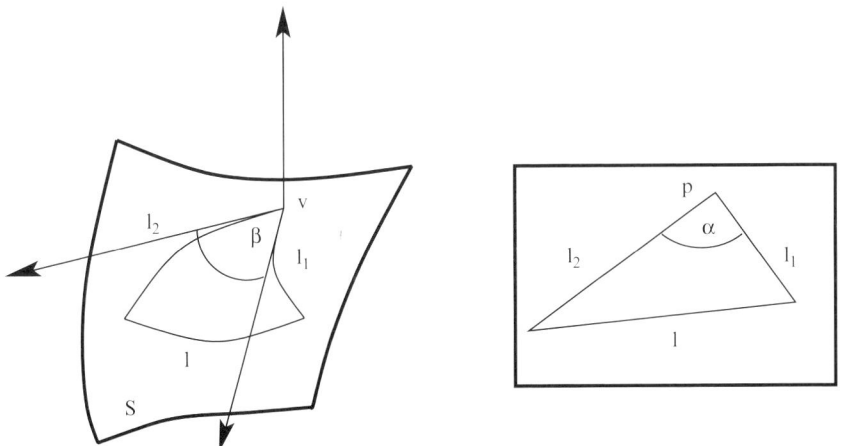

Fig. 3.8 The Euclidean triangle associated to a geodesic triangle on a surface S. The curvature at v is related to the angle at p and the angle at v by (3.5)

---

[2] In this formula, o denotes the Landau notation: $o(\bar{l}^2) = \bar{l}^2 \varepsilon(\bar{l})$, where ε is a function tending to 0 when $\bar{l}$ tends to 0.

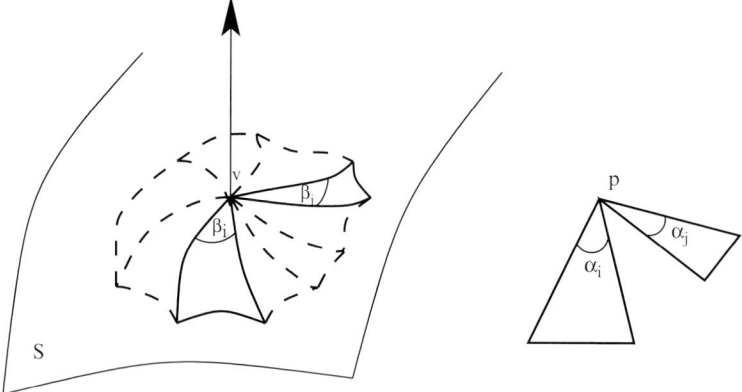

Fig. 3.9 Evaluating the Gauss curvature of a surface at a vertex v of a geodesic triangulation using (3.5)

Fig. 3.10 The α_i are the angles of the triangles incident to the vertex v

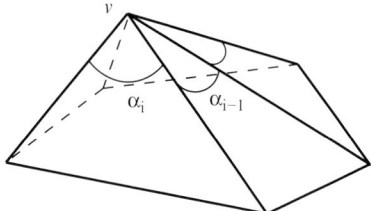

3.2.3 The Angular Defect of a Vertex of a Polyhedron

Let us now discuss a discrete analogue of the (smooth) Gauss curvature for polyhedra.

Looking at (3.6), it may be tempting to relate the Gauss curvature at a vertex of a polyhedron to its *angular defect* (up to a constant of normalization). We give a precise definition.

Definition 7. Let v be a vertex of a polyhedron P. The angular defect of P at v is the real number

$$\alpha_v = 2\pi - \sum_i \alpha_i,$$

where the α_i denote the angles at v of the triangles incident to P (Fig. 3.10).

Consequently, α_v is null if and only if the 1-ring of triangles around v is developable (Fig. 3.11).[3]

We shall improve this point of view in Chap. 6.

---

[3] Here, the 1-*ring of triangles around* v means the set of triangles which have v as a commun vertex, as we will see in Sect. 3.2.4.

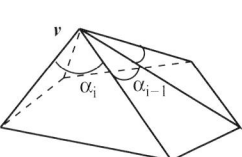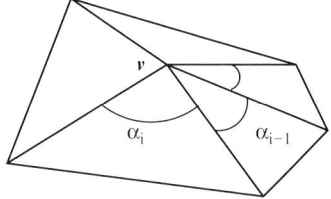

Fig. 3.11 If the sum of the α_i equals 2π, then v is developable, i.e., the 1-ring around v can be "flattened" onto a plane

Fig. 3.12 Approximating the Gauss curvature at a point p of a surface by the solid angle of the approximating triangulation at the vertex p gives in general bad results

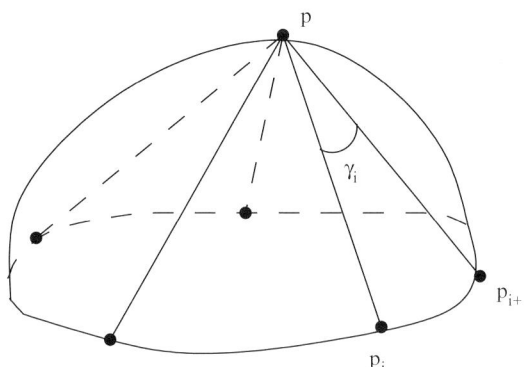

Note that although the Gauss curvature at a point of a smooth surface is homogeneous to the inverse of a surface area, the angular defect of a vertex of a polyhedron is dimensionless.

3.2.4 Warning! A Negative Result

Theorem 7 is of little help from a practical standpoint since the knowledge of geodesics is required. This leads us to replace the length of the geodesics in (3.6) by the length of their underlying chords, easy to evaluate, and the angles β_i by the angles between the corresponding chords, still easy to evaluate. However, the result is bad in full generality (Fig. 3.12).

A deep study of this phenomenon can be found in [20, 21]. We summarize here the main results, by giving a Taylor expansion of the angular defect in terms of the geometry of the surface around one of its points.

Consider a point p on a smooth surface S together with n Euclidean triangles $\{p_i p p_{i+1}\}$ (with $p_{n+1} = p_1$), forming a piecewise linear approximation of S around p. We introduce the following notation:

- The 1-*ring* around p is the set of points $p_1, ..., p_n$ which have a commun edge with p.

- The integer n is called the *valence* of p.
- The plane containing the normal to the surface, p, and p_i is denoted by Π_i; this plane is defined by its angle ϕ_i with respect to some coordinate system in the tangent plane of S at p (for instance, given by a frame defined by principal directions).
- The Euclidean distance from p to p_i is denoted by η_i.
- The angle at p of the triangle pp_ip_{i+1} is denoted by γ_i.
- The angle between two consecutive normal sections Π_i and Π_{i+1} is denoted by β_i.
- To simplify the notations, we put

$$c_i = \cos\phi_i, \; s_i = \sin\phi_i.$$

Now we put

$$A_i = \frac{1}{4\sin\gamma_i}[\eta_i\eta_{i+1}(c_i^2 s_{i+1}^2 + s_i^2 c_{i+1}^2) - \frac{\cos\gamma_i}{2}(\eta_i^2(2c_i^2 s_i^2) + \eta_{i+1}^2(2c_{i+1}^2 s_{i+1}^2))],$$

$$B_i = \frac{1}{4\sin\gamma_i}[\eta_i\eta_{i+1}(c_i^2 c_{i+1}^2) - \frac{\cos\gamma_i}{2}(\eta_i^2 c_i^4 + \eta_{i+1}^2 c_{i+1}^4)],$$

$$C_i = \frac{1}{4\sin\gamma_i}[\eta_i\eta_{i+1}(s_i^2 s_{i+1}^2) - \frac{\cos\gamma_i}{2}(\eta_i^2 s_i^4 + \eta_{i+1}^2 s_{i+1}^4)].$$

$$(3.7)$$

Let

$$A = \sum_i A_i, \; B = \sum_i B_i, \; C = \sum_i C_i.$$

One has the following theorem.

Theorem 8. *At each point p of S,*

$$2\pi - \sum_{i=1}^{n}\gamma_i = (AG_p + B\lambda_1^2 + C\lambda_2^2) + o(\eta^2),$$

$$(3.8)$$

where λ_1 and λ_2 denote the two principal curvatures of S at p.

In full generality, Theorem 8 shows that there is no hope of getting any interesting approximation of the Gauss curvature of the surface, up to order 2, with the solid angle of the 1-ring neighbor given by the chords.

The only possibility to improve (3.8) is to find suitable triangulations around p which cancel the terms B and C. This happens for triangulations having some kind of regularity. That is why we give the following definition.

Definition 8. Let p be a point of a smooth surface S and $p_i, i = 1, ..., n$ be its 1-ring neighbors. The point p is called a regular vertex if:

- The points p_i lie in two consecutive normal sections of which form an angle $\theta = \frac{2\pi}{n}$.
- The lengths η_i take the same value η.

Using these particular triangulations, we get the following theorem.

Theorem 9. *Consider a regular vertex p with valence* 6. *Then*

$$\lim_{\eta \to 0} \frac{2\pi - \sum_i \gamma_i}{\eta^2} = \frac{\sqrt{3}}{2} G_p.$$

3.2.5 Warning! The Pointwise Gauss Curvature of a Closed Surface

It is well known that a C^∞ closed[4] surface S in $\mathbb{E}^3$ has a point p with positive Gauss curvature. This result fails for closed polyhedra. For instance, there exist embedded flat polyhedric tori T in $\mathbb{E}^3$: each vertex v of T has a null angular defect ($\alpha_v = 0$) (Figs. 3.13 and 3.14).

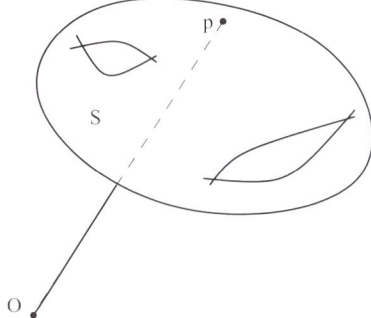

Fig. 3.13 The point p of S which is at a maximal distance from 0 has a positive Gauss curvature

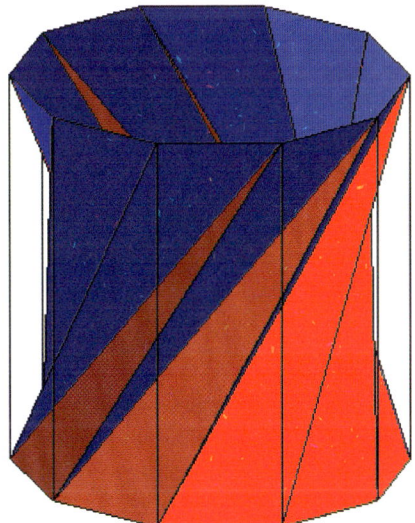

Fig. 3.14 Here is a flat polyhedric torus embedded in $\mathbb{E}^3$: each vertex has a null angular defect. This torus has 16 "internal" faces (*triangles*), and 8 external faces (*transparent rectangles*)

---

[4] That is, compact without boundary.

Such a torus can be constructed as follows: it is the union of n triangles

$$(A_k, A_k, B_{k+3}),$$

n triangles

$$(B_k, B_{k-1}, A_{k-3}),$$

and n rectangles

$$(A_k, A_{k+1}, B_{k+1}, B_k),$$

where

$$A_k = \left(\cos\frac{2k\pi}{n}, \sin\frac{2k\pi}{n}, -1\right), \, B_k = \left(\cos\frac{2k\pi}{n}, \sin\frac{2k\pi}{n}, 1\right).$$

When $n \geq 7$, T is embedded in $\mathbb{E}^3$.[5]

3.2.6 Warning! A Negative Result Concerning the Approximation by Quadrics

We have seen in Sect. 3.2.4 that there is in general no hope of getting a good approximation to the curvature at a point p of a curve γ, by taking four points in general position on γ close to p, the conic $\mathcal{C}$ containing p and these four points, and computing the curvature of $\mathcal{C}$ at p. In the same spirit, there is in general no hope to get a good approximation to the curvatures at a point p of a surface S, by taking nine points in general position on γ close to p, using the quadric $\mathcal{Q}$ through p and these nine points, and computing the curvatures of $\mathcal{Q}$ at p.

More generally, the following situation shows the impossibility of finding "universal" discrete geometric pointwise quantities to approximate the curvatures of a surface:

- Consider two surfaces S_1 and S_2 whose intersection is the union of two secant smooth curves γ_1 and γ_2. Let $p = \gamma_1 \cap \gamma_2$.
- Suppose that the Gauss curvature G_{1_p} of S_1 is different to the Gauss curvature G_{2_p} of S_2 at p.
- Now, consider four sequences of points q_n, r_n, s_n, t_n such that $q_n p r_n$ are on γ_1 (in this order) and $s_n p t_n$ are on γ_2 (in this order), tending to p when n tends to infinity.
- Let T_n be the triangulated surface whose faces are the triangles

$$pqs, psr, prt, pqt.$$

Associate to this piecewise linear surface any number ζ_n, called "the curvature" of T_n. When n tends to infinity, ζ_n cannot tend to both G_{1_p} and G_{1_p}, since these quantities are different by construction. The same argument can be invoked with the mean curvatures or any principal curvature (Fig. 3.15).

This point of view has been extensively developed and studied in [84].

---

[5] The author thanks R. Ferréol and G. Valette for interesting private mails on this example.

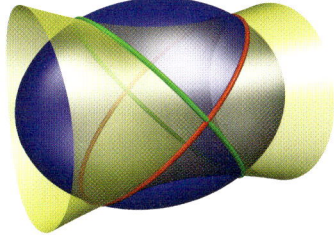

Fig. 3.15 Here is the intersection of a hyperboloid (of negative Gauss curvature) and an ellipsoid (of positive Gauss curvature). This intersection is the union of two curves. By constructing a sequence of triangulations inscribed in these curves and tending to one of their intersection point, $(T_n)_{n \in \mathbb{N}}$, one sees that any sequence of number ζ_n can converge to both the Gauss curvature of the hyperboloid and the Gauss curvature of the ellipsoid. This image is courtesy of the I.N.R.I.A. project team Vegas

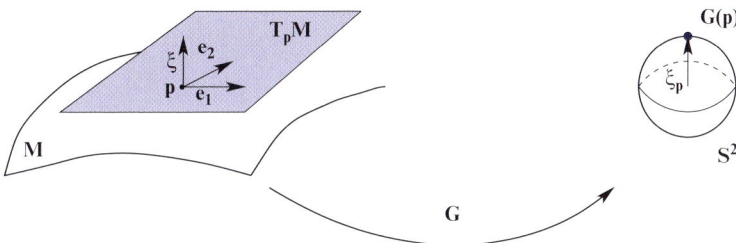

Fig. 3.16 The Gauss map of a surface

3.3 The Gauss Map of a Surface

3.3.1 The Gauss Map of a Smooth Surface

Let S be an oriented (smooth) surface of $\mathbb{E}^3$. The Gauss map associated to S is the map (Fig. 3.16)

$$\mathbf{G} : S \to \mathbb{S}^2,$$

defined as follows: if ξ is the oriented normal vector field of S, then for all $m \in \mathbb{S}^2$

$$\mathbf{G}(m) = \xi_m.$$

The relation between the Gauss map $\mathbf{G}$ and the Gauss curvature of S can be seen as follows: if $da_{\mathbb{S}^2}$ is the area form of $\mathbb{S}^2$, one has

$$\mathbf{G}^*(da_{\mathbb{S}^2}) = G\,da_S, \tag{3.9}$$

where da_S denotes here the area form of S and, as usual, G is the Gauss curvature function of S (see [11, 40] for instance). Consequently, integrating (3.9) gives a geometrical interpretation of the total Gauss curvature in terms of an algebraic area:

$$\int_S \mathbf{G}^*(da_{\mathbb{S}^2}) = \int_S G \, da_S. \tag{3.10}$$

If the map G is one to one, (3.10) implies that

$$\int_S G \, da_S = \pm \int_{\text{im}\,(\mathbf{G})} da_{\mathbb{S}^2}, \tag{3.11}$$

the sign depending on the orientation.

If the map G is not one to one, the interpretation of (3.9) requires care, since some parts of the image of G are covered many times, introducing an index in (3.11) to recover a correct equality.

3.3.2 The Gauss Map of a Polyhedron

One can generalize the concept of the Gauss map of smooth surfaces to polyhedra (see [8]). If P is an (oriented) polyhedron of $\mathbb{E}^3$, the Gauss map associates to each vertex v of P its exterior angle,[6] considered as an oriented subset $\mathbf{a}_v$ of $\mathbb{S}^2$ endowed with an index. This index describes how often a_v is covered. For instance, if v is a convex vertex with respect to P, then this index is 1. For a vertex of saddle type, the index is -1, and for a monkey saddle, this index is -2. Figures 3.17–3.19 reproduce the standard cases studied in [8] (we deal with normal *vectors* instead of normal *lines* appearing in [8]).

The interest of this point of view is that it allows Gauss–Bonnet theorem to be recovered in terms of critical points of height functions in a broad frame including smooth surfaces and polyhedra.

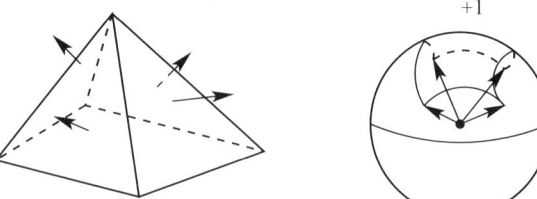

Fig. 3.17 If v is a convex vertex of a polyhedron, the index of $\mathbf{a}_v = 1$

---

[6] A precise definition is given in Chap. 6.

Fig. 3.18 If v is a saddle point, the index of $\mathbf{a}_v = -1$

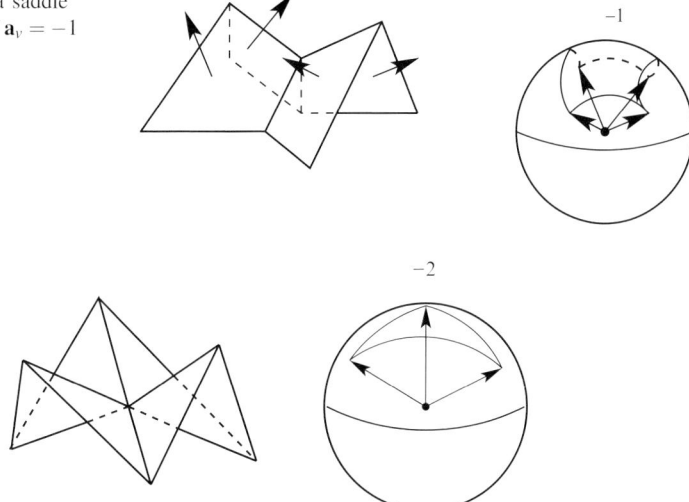

Fig. 3.19 If v is the vertex of a monkey saddle, each of the three vectors drawn in $\mathbb{S}^2$ is the (unit) normal vector of two faces of the monkey saddle. Taking the orientation in account leads to $\mathbf{a}_v = -2$

3.4 The Global Gauss Curvature

If S is a closed (oriented) surface of $\mathbb{E}^3$, we deduce from (3.9) that its global curvature is given by the following equality:

$$\mathcal{K}(S) = \int_S \mathbf{G}^*(da_{\mathbb{S}^2}).\tag{3.12}$$

A remarkable relationship relating the global Gauss curvature of a (smooth) closed surface S to its topology is the Gauss–Bonnet theorem (see [11, 40] for instance): the Euler characteristic $\chi(S)$ of the surface[7] S can be evaluated by integrating the Gauss curvature function G of S. One has

$$\mathcal{K}(S) = 2\pi\chi(S).\tag{3.13}$$

This smooth theorem has a discrete analogue: if P is a polyhedron, its Euler characteristic $\chi(P)$ can be evaluated by summing the Gauss curvature of P at each vertex [8]:

$$2\pi\chi(P) = \sum_{v \text{ vertex of } P} \alpha_v,\tag{3.14}$$

where α_v denotes the *angular defect* of P at v.

---

[7] We assume here that S has no boundary, but an equivalent result is valid for general oriented surfaces (with or without boundary).

Fig. 3.20 The area of the smooth surface S of $\mathbb{E}^2$ is approximated by the area of the polyhedron P (or Q)

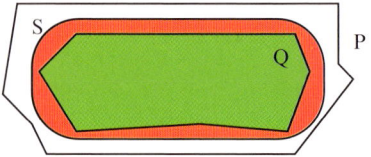

By analogy with the smooth case, it is then natural to define the (global) Gauss curvature of P by setting

$$\mathcal{K}(P) = \sum_{v \text{ vertices in } P} \alpha_v. \tag{3.15}$$

This point of view will be generalized to polyhedra of any dimension in Chap. 6.

In our context (of approximation), these fundamental results appear as negative ones. In fact, this means that the shape of a smooth surface or a polyhedron cannot be detected by integrating or summing the pointwise Gauss curvature. Deforming a (smooth or discrete) surface does not change its topology and does not affect its global Gauss curvature.

3.5 The Volume

Although the volume functional is not continuous on the class of subsets of $\mathbb{E}^3$ for the Hausdorff topology, it is interesting to point out that its additivity property implies simple interesting results concerning its approximation. For instance, if a three-dimensional (bounded) domain D of $\mathbb{E}^3$ lies between two (bounded) domains P_1 and P_2:

$$P_1 \subset D \subset P_2,$$

then the volume of D is bounded from below by the volume of P_1 and from above by the volume of P_2. If the difference of the volumes of P_1 and P_2 is "small," then a "good" approximation of the volume of D is given by the volume of P_1 or P_2. This remark gives a simple way to approximate the volume of a domain (with smooth boundary) by volumes of polyhedra (Fig. 3.20).

The Aim of This Book

The aim of this book is now clear: to construct a theory which allows us to define geometric invariants (which will be called *curvature measures*) on a large class of objects including smooth submanifolds, piecewise linear spaces, convex subsets (and more generally, singular objects). By using a coherent topology, our goal is to derive approximation and convergence results.

Part II
Background: Metrics and Measures

Chapter 4
Distance and Projection

The objects studied here are subsets of the Euclidean space $\mathbb{E}^N$, endowed with its classical scalar product $< .,. >$. If

$$x = (x_1, ..., x_i, ..., x_N), y = (y_1, ..., y_i, ..., y_N)$$

are two points of $\mathbb{E}^N$, then

$$< x, y > = \sum_{i=1}^{N} x_i y_i.$$

This scalar product induces a norm on $\mathbb{E}^N$

$$|x| = \sqrt{\sum_{i=1}^{N} x_i^2}$$

and a metric d on $\mathbb{E}^N$, defined by

$$d(x, y) = |x - y| = |\overrightarrow{xy}|.$$

As usual, if x is a point of $\mathbb{E}^N$ and $r > 0$ is a real number, the *open ball* $B(x, r)$ is the set of points whose distance to x is strictly smaller than r and the *closed ball* $\bar{B}(x, r)$ is the closure of $B(x, r)$, i.e., the set of points whose distance to x is less than or equal to r.

4.1 The Distance Function

The distance d on $\mathbb{E}^N$ can be extended as follows. If m is a point of $\mathbb{E}^N$ and A is a (nonempty) subset of $\mathbb{E}^N$, one defines the distance $\mathrm{d}(m, A)$ from m to A by

$$d(m, A) = \inf_{a \in A} d(m, a).$$

Fig. 4.1 The tube of radius ε of a "full" rectangle A in $\mathbb{E}^2$

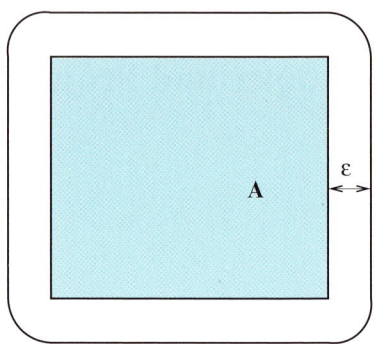

Fig. 4.2 The tube of radius ε of an "empty" rectangle R in $\mathbb{E}^2$ is the set of points which are between the two boldface lines

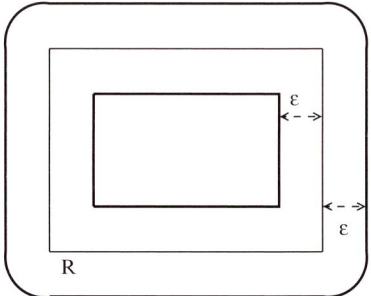

Now, fixing A, one can introduce the distance function d_A as the map

$$d_A : \mathbb{E}^N \to \mathbb{R}^+$$

by setting, for all $m \in \mathbb{E}^N$,

$$d_A(m) = d(m,A).$$

We define now the *tube* around a subset A (Figs. 4.1 and 4.2).

Definition 9. Let A be any subset of $\mathbb{E}^N$ and $\varepsilon > 0$. The tube or tubular neighborhood of A is the set of points

$$A_\varepsilon = \{m \in \mathbb{E}^N, d(m,A) \leq \varepsilon\}.$$

Using the notion of tube, one defines the most common distance between two (nonempty) subsets of $\mathbb{E}^N$, namely the *Hausdorff distance*.

Let $\mathcal{P}(\mathbb{E}^N)$ denote the set of all subsets of $\mathbb{E}^N$.

Definition 10. The *Hausdorff distance* d is defined as follows: for all elements P and Q of $\mathcal{P}(\mathbb{E}^N)$,

$$d(P,Q) = \inf\{r > 0 : P \subset Q_r \text{ and } Q \subset P_r\}. \tag{4.1}$$

Fig. 4.3 The length l (resp., l') denotes the distance from m to A (resp., from l' to m'). One has always $|l - l'| \leq |mm'|$

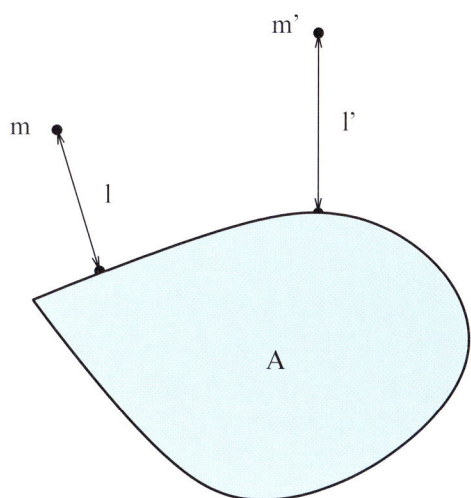

One can easily check that d is effectively a distance on the set $\mathcal{P}_c(\mathbb{E}^N)$ of compact subsets of $\mathbb{E}^N$.

The main properties of d_A are summarized in the following theorem.

Theorem 10. *Let A be a (nonempty) subset of $\mathbb{E}^N$. Then:*

1. $d_A(m) = 0 \iff m \in \overline{A}$.
2. The map

$$d_A : \mathbb{E}^N \rightarrow \mathbb{R}^+$$

is uniformly Lipschitz and

$$\forall m, m' \in \mathbb{E}^N, d_A(m) - d_A(m') \leq d(m, m'). \tag{4.2}$$

3. The map d_A is differentiable almost everywhere and

$$|\mathrm{grad}\, d_A| = 1, \tag{4.3}$$

almost everywhere.

The proof of Theorem 10 is classic and left to the reader (see [37] for instance). We shall improve the equality (4.3) by using the projection map in Theorem 12 (Fig. 4.3).

4.2 The Projection Map

If m is a point of $\mathbb{E}^N$, one can consider the subset of points $a \in A$ such that $d_A(m) = d(m, a)$. If A is closed, this subset is not empty. Any point of it is called an *orthogonal projection* of m onto A (Fig. 4.4).

Fig. 4.4 The point m has
three orthogonal projections
onto the curve C

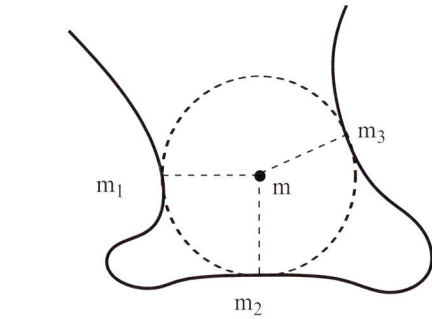

Fig. 4.5 When A is smooth,
the line $\mathrm{pr}(m)m$ is orthogonal
to the tangent space $T_{\mathrm{pr}(m)}A$

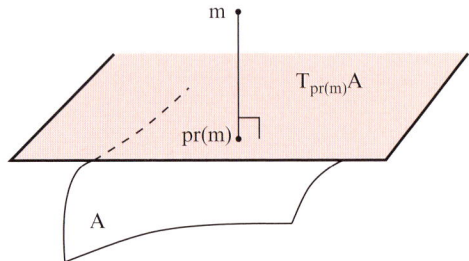

Let us denote by U_A the subset of points of $\mathbb{E}^N$ having a *unique* orthogonal projection onto A. On U_A, one can define the orthogonal projection map

$$\mathrm{pr}_A : U_A \to A$$

(if there is no possible confusion, one simply writes pr instead of pr_A).

Remarks

- Since d_A is Lipschitz, it is differentiable almost everywhere. Using Theorem 10, this implies that pr_A is defined almost everywhere and $\mathbb{E}^N \setminus U_A$ has a null (Lebesgue) measure.
- Moreover, if A is smooth, then $\overrightarrow{\mathrm{pr}_A(m)m}$ is orthogonal to A, i.e., $\overrightarrow{\mathrm{pr}_A(m)m}$ is orthogonal to the tangent space $T_{\mathrm{pr}_A(m)}A$ of A at $\mathrm{pr}_A(m)$ (Fig. 4.5).

In general, the projection map is not continuous, as the example of Fig. 4.6 shows. However, by restricting the geometry or the topology of the subset A, one may obtain continuity. For instance, one has the following theorem.

Theorem 11. *Let A be a compact subset of $\mathbb{E}^N$. Then, the orthogonal projection*

$$\mathrm{pr} : U_A \to A$$

is continuous.

Fig. 4.6 When it is defined, the map pr_A is not continuous in general: here, A is the union of two intervals. When m tends to m_0, $\mathrm{pr}_A(m)$ does not tend to $\mathrm{pr}_A(m_0)$

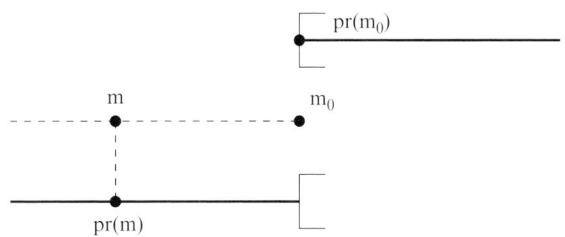

Sketch of proof of Theorem 11 Let $m \in U_A$ and let p be the projection of m onto A. Let $(m_n)_{n \in \mathbb{N}}$ be a sequence of points of U_A which tends to m. For all $n \in \mathbb{N}$, let p_n be the orthogonal projections of m_n onto A. We shall prove that

$$\lim_{n \to \infty} p_n = p.$$

Since A is compact, there exists a subsequence of $(p_n)_{n \in \mathbb{N}}$, which we still denote by (p_n), $n \in \mathbb{N}$, which converges to a point q of A. Since the distance function d_A is continuous, $d_A(m_n)$ tends to $d_A(m) = d(m, p)$. On the other hand,

$$d_A(m_n) = d(m_n, A) = d(m_n, p_n)$$

tends to $d(m, q)$. Hence,

$$d(m, p) = d(m, q).$$

Since m has a unique projection on A, $p = q$ and

$$\lim_{n \to \infty} p_n = p.$$

This implies that all limiting values of the sequence $(p_n)_{n \in \mathbb{N}}$ equal p. Classically, we conclude that the sequence $(p_n)_{n \in \mathbb{N}}$ tends to p. This implies that the orthogonal projection is continuous on U_A. $\square$

The main properties of the distance function are summarized in the following theorem.

Theorem 12. *Let A be a (nonempty) closed subset of $\mathbb{E}^N$ and let m be a point of $\mathbb{E}^N$:*

1. If grad $d_A(m)$ *exists, then m lies in U_A.*
2. If m is a point of A such that grad $d_A(m)$ *exists, then*

$$\mathrm{grad}\ d_A(m) = 0.$$

3. For all $m \notin A$,

$$\mathrm{grad}\ d_A(m) = \frac{\overrightarrow{\mathrm{pr}_A(m)m}}{d_A(m)}.$$

4. d_A is C^1 on int$(U_A \setminus A)$.
5. d_A^2 is continuously differentiable on int U_A and

$$\mathrm{grad}\ d_A^2(m) = 2d(m, \mathrm{pr}(m))\mathrm{grad}\ d_A(m).$$

Sketch of proof of Theorem 12

1. Let $m \in (\mathbb{E}^N \setminus A)$. Suppose that the function d_A is differentiable at m and that there exist two different points $a, a' \in A$ such that

$$d_A(m) = d(m,a) = d(m,a').$$

The kernel of the differential $Dd_A(m)$ is orthogonal to the vectors ma and ma', and its dimension is $(N-1)$, which is a contradiction. Consequently, there exists only one point a satisfying $d_A(m) = d(m,a)$ and $m \in U_A$.

2. Let m be a point of A such that $\operatorname{grad} d_A(m)$ exists. This point is clearly a minimum for the function d_A, which implies that its gradient is null at m.

3. Let m be a point which does not belong to A. Suppose that $\operatorname{grad} d_A(m)$ exists. The gradient $\operatorname{grad} d_A(m)$ of d_A at m is parallel to the vector $\overrightarrow{ma}$, where a denotes the (orthogonal) projection of m onto A. On the other hand, since

$$|d_A(m) - d_A(m')| \le d(m,m'),$$

we have $|Dd_A(m)| \le 1$. Moreover,

$$d_A[m + t(a-m)] = d_A(m) - t d_A(m).$$

We deduce that

$$\operatorname{grad} d_A(m) = \frac{\overrightarrow{\operatorname{pr}_A(m)m}}{d_A(m)}.$$

4. The last two items are easy, well known, and left to the reader. □

4.3 The Reach of a Subset

Following Federer [42], we give the following definition.

Definition 11.

1. Let A be a subset of $\mathbb{E}^N$ and $a \in A$. We denote by $\operatorname{reach}(A,a)$ the supremum of real numbers r such that every point of the ball of radius r and center a has a unique projection onto A (realizing the distance $d_A(m)$ from m to A):

$$\operatorname{reach}(A,a) = \sup\{r \in \mathbb{R} \text{ for which } \{m : d(m,a) < r\} \subset U_A\}.$$

2. The reach of A is the real number

$$\operatorname{reach}(A) = \inf_{a \in A} \operatorname{reach}(A,a).$$

In other words, $\operatorname{reach}(A)$ is the supremum of real numbers $r \ge 0$ such that the orthogonal projection onto A is defined in the tubular neighborhood of radius r.

For instance, the reach of a circle C of radius r is r since every point of $\mathbb{E}^2$ except the center O has a unique orthogonal projection onto C (Fig. 4.7).

Remarks

- If γ is a smooth curve in $\mathbb{E}^2$ and $x \in \gamma$, then for all $a \in A$,

$$k_a \text{reach}(A, a) \le 1, \tag{4.4}$$

  where k_a is the curvature of γ at the point a.
- If S is a smooth surface in $\mathbb{E}^3$ and $x \in S$, then for all $a \in A$,

$$|h|_a \text{reach}(A, a) \le 1, \tag{4.5}$$

  where $|h|_a$ denotes the norm of the second fundamental form of S at a.
- The example given in Fig. 4.8 shows that the previous inequalities may be strict. The reason is that the curvatures of smooth objects are local geometric invariants. However, the reach is both a local and a global invariant.
- Of course, convex sets have an infinite reach (Fig. 4.9).
- Conversely, the reach of a polygonal line with more than one edge is null (Fig. 4.10).
- If the reach of a subset A is r, then every point belonging to the tube of A of radius r has a unique projection onto A (realizing the distance from a to A) (Fig. 4.11).

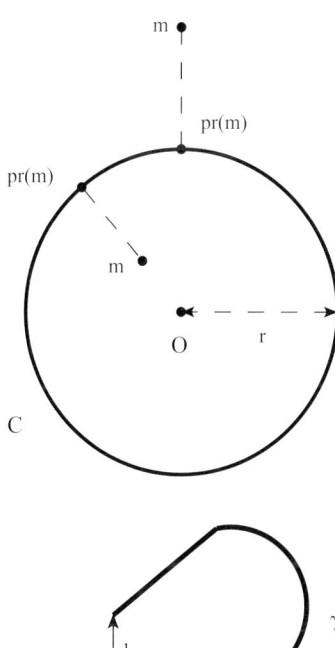

Fig. 4.7 The reach of a circle of radius r is r

Fig. 4.8 The reach of the curve γ at the point a is $\frac{l}{2}$. It is smaller than the inverse of the curvature of γ at this point, since the curvature at a is 0

Fig. 4.9 Any point in $\mathbb{E}^N$ has a unique projection on the convex subset C; consequently, the reach of C is $+\infty$

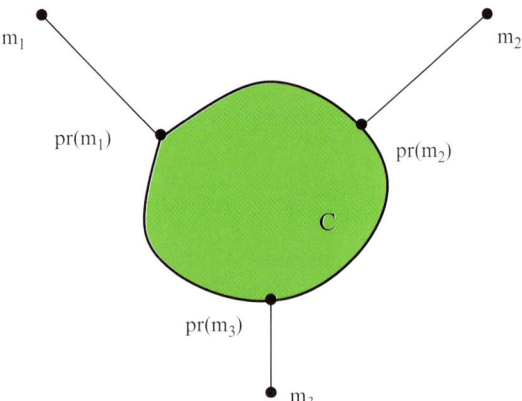

Fig. 4.10 The reach of this polygonal line P in $\mathbb{E}^2$ is null since every point of the bisector has two orthogonal projections on P

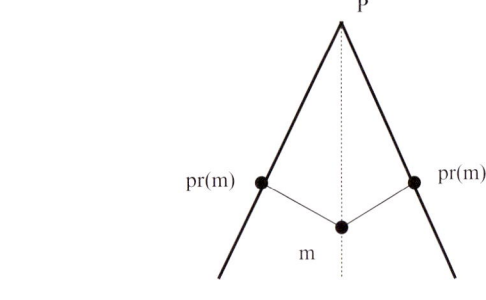

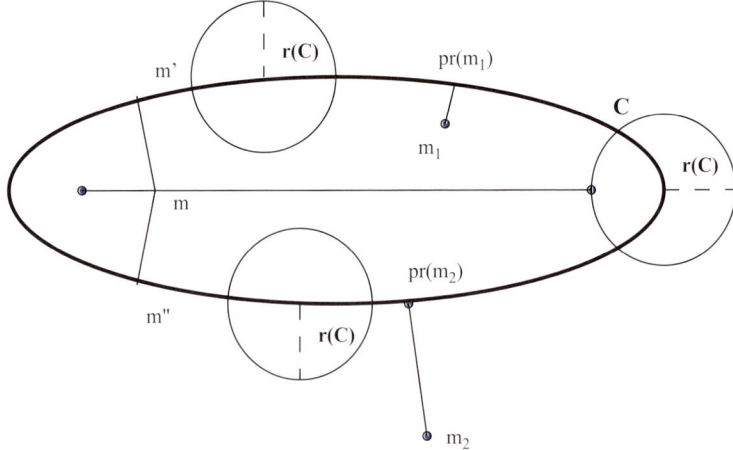

Fig. 4.11 If C is an ellipse in the plane, $U_C = \mathbb{E}^2 \setminus D$ is a tubular neighborhood of C, where D is the segment joining the two foci. The reach $r(C)$ of C is the distance between one focus and C

Fig. 4.12 A Voronoi diagram
in $\mathbb{E}^2$

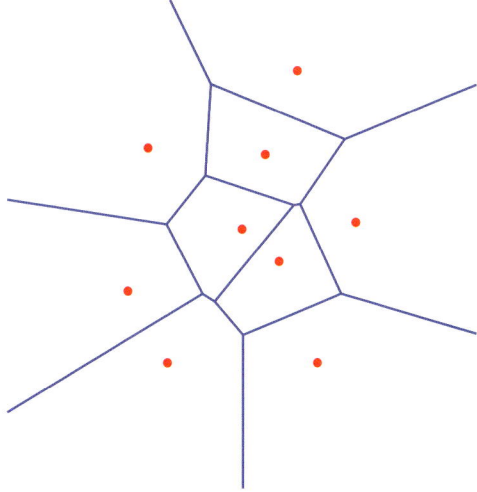

4.4 The Voronoi Diagrams

If A is a finite subset of $\mathbb{E}^N$, one can associate to A a partition of $\mathbb{E}^N$, formed with help of the distance function. We summarize this construction. Suppose that $A = \{p_1, \ldots, p_k\}$. To each p_i, we associate its *Voronoi region* $\mathrm{Vor}(p_i)$, i.e.,

$$\mathrm{Vor}(p_i) = \{x \in \mathbb{E}^N : |x - p_i| \leq |x - p_j|, \forall j \leq k\}.$$

Here, $\mathrm{Vor}(p_i)$ is the intersection of the $k - 1$ half-spaces bounded by the bisector planes of p_i and each of the other points of A. Note that $\mathrm{Vor}(p_i)$ is a convex polytope.

Definition 12. The Voronoi diagram $\mathrm{Vor}(A)$ of A is the cell complex whose N-cells are the Voronoi regions (Fig. 4.12).

4.5 The Medial Axis of a Subset

Two important tools in surface reconstruction are the *medial axis* and the *local feature size*, much related to the reach. We can define them in full generality.[1]

Definition 13. Let A be a subset of $\mathbb{E}^N$. The *medial axis* $\mathrm{Med}(A)$ of A is the closure of the set of points which have at least two closest points on A.

Definition 14. The local feature size $\mathrm{lfs}(a)$ at a point $a \in A$ is the distance of a from the medial axis of A.

---

[1] There is no uniform definition of these notions in the literature. We give here the simplest one.

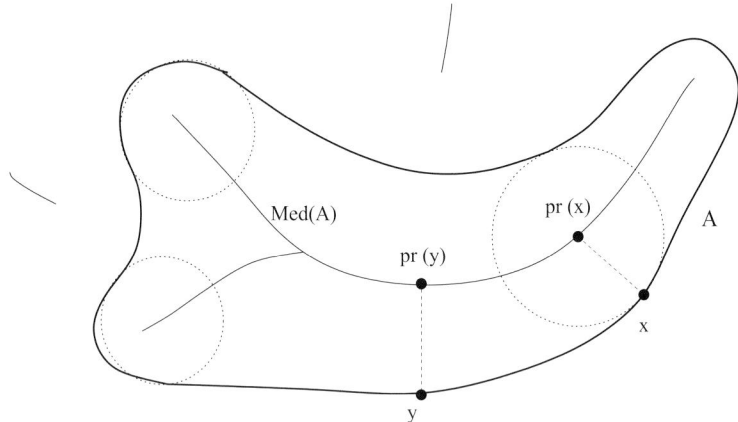

Fig. 4.13 Any point on the medial axis has at least two orthogonal projections. One has $\mathrm{lfs}(x) = |x - \mathrm{pr}(x)|, \mathrm{lfs}(y) = |y - \mathrm{pr}(y)|$

In other words, using the notion of *reach* introduced above, one has

$$\mathrm{lfs}(a) = \mathrm{reach}(A, a).$$

Note that if A is finite, the medial axis of A is nothing but the boundary of the Voronoi regions associated to A (Fig. 4.13).

Using the triangle inequality, one proves immediately the following theorem.

Theorem 13. *Let A be a subset of $\mathbb{E}^N$, let $a \in A$, and let x and y be two points of A. Then*

$$|\mathrm{lfs}(x) - \mathrm{lfs}(y)| \leq |x - y|.$$

Indeed,

$$\mathrm{lfs}(x) = |x - \mathrm{pr}(x)| \leq |x - \mathrm{pr}(y)| \leq |x - y| + |y - \mathrm{pr}(y)|,$$

from which Theorem 13 follows.

Chapter 5
Elements of Measure Theory

The goal of this book is to associate to a large class of subsets of $\mathbb{E}^N$ a *signed measure* which allows us to evaluate their geometry. In many cases (convex subsets and smooth subsets), one tool is the computation of the *volume* of tubular neighborhoods. That is why *measure theory* is indispensable in our context. We summarize in this chapter the main constructions and results on *outer measures*, *measures*, and *signed measures* ([43,53,63] are the basic references).

5.1 Outer Measures and Measures

Although we deal with $\mathbb{E}^N$, we begin in a very large framework, defining *measures* on any set X. Let $\mathcal{P}(X)$ be the collection of subsets of X.

5.1.1 Outer Measures

An *outer measure* μ on the set X is a map

$$\mu : \mathcal{P}(X) \to \overline{\mathbb{R}}^+$$

such that:

- The empty set $\emptyset$ has measure 0:

$$\mu(\emptyset) = 0.$$

- If $A \subset B$, then

$$\mu(A) \le \mu(B). \tag{5.1}$$

- μ is countably subadditive, i.e., it satisfies, for any countable family of disjoint elements of $\mathcal{P}(X)$,

$$\mu(\cup_{n\in\mathbb{N}}A_n) \leq \sum_{n\in\mathbb{N}} \mu(A_n). \tag{5.2}$$

5.1.2 Measures

- Let X be a set. A *σ-algebra* $\mathcal{A}$ of X is a family of subsets of X, which contains the empty set $\emptyset$ and which is closed under countable unions and complements. This implies in particular that $\mathcal{A}$ is also closed under countable intersections and differences. The couple $(X, \mathcal{A})$ is called a *measurable space*.
- A *measure* μ on the σ-algebra $\mathcal{A}$ is a map

$$\mu : \mathcal{A} \to \overline{\mathbb{R}}^+$$

such that:

- The empty set $\emptyset$ has measure 0:

$$\mu(\emptyset) = 0.$$

- μ is countably additive, i.e., it satisfies, for any countable family of disjoint elements of $\mathcal{A}$,

$$\mu(\cup_{n\in\mathbb{N}}A_n) = \sum_{n\in\mathbb{N}} \mu(A_n). \tag{5.3}$$

In particular, this implies that:

- For every A and B in $\mathcal{A}$,

$$\mu(A\cup B) = \mu(A) + \mu(B) - \mu(A\cap B). \tag{5.4}$$

- If $A \subset B$, then
$$\mu(A) \leq \mu(B). \tag{5.5}$$

We say that the triple $(X, \mathcal{A}, \mu)$ is a *measure space* and the elements of $\mathcal{A}$ are *μ-measurable*.

- Let P be a property on the point x of $(X, \mathcal{A}, \mu)$. We say that P is true *μ-almost everywhere* if there exists $A \in \mathcal{A}$ of null measure ($\mu(A) = 0$), such that the points x which do not satisfy the property P belong to A.

5.1.3 Outer Measures vs. Measures

There are two main differences between *outer measures* and *measures* on a set X:

1. An *outer measure* measures *any* subset of X, but is only countably subadditive.
2. A *measure* is countably additive, but it can only measure the subsets which belong to a σ-algebra of X.

Of course, the best situation is the case of a space X endowed with a measure defined on $\mathcal{P}(X)$. This is always possible (for instance, taking the null measure), but it becomes often impossible if one needs particular interesting properties of the measure, in particular in $\mathbb{E}^N$ (see Sect. 5.3). However, from any *outer measure*, one can define a *measure* on a suitable σ-algebra. This construction can be done as follows.

Let X be a set endowed with an outer measure μ^*. A subset A of X is μ^*-*measurable* if, for all $B \subset X$,

$$\mu^*(B) = \mu^*(B \cap A) + \mu^*(B \setminus A).$$

We denote by $\mathcal{M}_{\mu^*}$ the collection of all μ^*-measurable subsets of X.

The following result shows that "the restriction of μ^* to $\mathcal{M}_{\mu^*}$" is a measure. In particular, if A and B belong to $\mathcal{M}_{\mu^*}$,

$$\mu(A \cup B) = \mu(A) + \mu(B) - \mu(A \cap B). \tag{5.6}$$

Theorem 14. *Let X be a set endowed with an outer measure μ^*. Then:*

- *The collection $\mathcal{M}_{\mu^*}$ of all μ^*-measurable subsets of X is a σ-algebra.*
- *The restriction of μ^* to $\mathcal{M}_{\mu^*}$ is a measure on $\mathcal{M}_{\mu^*}$.*

5.1.4 Signed Measures

The previous context allows us to define the classical *N-volume* in $\mathbb{E}^N$ as a Borel measure on $\mathbb{E}^N$. The *curvature measures* introduced in this book are not strictly speaking measures, since they may be negative on some Borel subsets of $\mathbb{E}^N$. That is why we must generalize slightly the concept of measure, by introducing *signed measures*[1] on $\mathbb{E}^N$.

Definition 15. A signed measure μ on a σ-algebra $\mathcal{A}$ of X is a map

$$\mu : \mathcal{A} \to \mathbb{R},$$

which is countably additive, i.e., such that for any countable family of disjoint elements of $\mathcal{A}$:

- The empty set has measure 0:
$$\mu(\emptyset) = 0.$$
- The series $\sum_{n \in \mathbb{N}} \mu(A_n)$ converges.
-
$$\mu(\cup_{n \in \mathbb{N}} A_n) = \sum_{n \in \mathbb{N}} \mu(A_n). \tag{5.7}$$

Note that a bounded measure is a *signed* measure, but the converse is wrong in general.

---

[1] Some authors simply use the term *measure* instead of *signed measure*. This can be done if no confusion is possible.

5.1.5 Borel Measures

Suppose now that X is a topological space. The smallest σ-algebra $\mathcal{B}$ containing the open sets of X is called the *Borel σ-algebra* and its elements are called the *Borel sets*:

- An outer measure μ on X is said to be *Borel regular* if Borel sets are μ-measurable and every subset of X is contained in a Borel set of the same μ-measure.
- Any measure defined on the Borel σ-algebra $\mathcal{B}$ is called a *Borel measure*.

5.2 Measurable Functions and Their Integrals

5.2.1 Measurable Functions

- Let $(X, \mathcal{A})$ be a measurable space. A function

$$f : X \to \mathbb{R}$$

is *measurable* if the inverse image by f of any open subset of $\mathbb{R}$ belongs to $\mathcal{A}$. In particular, if A is a measurable subset of X, the characteristic function χ_A of A, defined by

$$\chi_A(x) = \begin{cases} 1 & \text{if } x \in A, \\ 0 & \text{if } x \notin A, \end{cases}$$

is measurable.

Remark The notion of *measurable function* can be extended to the case of a set X endowed with an outer measure μ. The definition is modified as follows: a function

$$f : X \to \mathbb{R}$$

is *measurable* if the inverse image by f of any open subset of $\mathbb{R}$ is a μ-measurable subset of (X, μ).

In the following, we shall essentially deal with measurable functions on a measure space $(X, \mathcal{A}, \mu)$. The reader will adapt the results to the case of a set endowed with an outer measure.

- We say that a measurable function f on $(X, \mathcal{A}, \mu)$ is *simple* if it takes only a finite number of values $(\alpha_i, 1 \le i \le n)$. In this case, $f^{-1}(\alpha_i)(1 \le i \le n)$ belongs to $\mathcal{A}$. Conversely, if $(A_1, ..., A_n)$ is a (finite) sequence of $\mathcal{A}$ and $(\alpha_1, ..., \alpha_n)$ is a sequence of real numbers, one can define the *simple* measurable function

$$s = \sum_1^n \alpha_i \chi_{A_i}.$$

We denote by $\mathcal{S}$ (resp., $\mathcal{S}^+$) the set of all simple functions (resp., the set of non-negative simple functions) on X.

Let f be a nonnegative measurable function on $(X, \mathcal{A}, \mu)$. Then, it can be proved that there exists an increasing sequence $(s_n)_{n \in \mathbb{N}}$ of nonnegative simple measurable functions such that

$$\forall x \in X, \lim_{n \to \infty} s_n(x) = f(x).$$

5.2.2 Integral of Measurable Functions

We can now define the integral of a nonnegative measurable function. We shall restrict our attention to a measure space $(X, \mathcal{A}, \mu)$; the reader will adapt the definition to a set endowed with an outer measure. We begin with the integral of a simple nonnegative function.

- Let $(A_i \in \mathcal{A}, 1 \leq i \leq n)$ and $(\alpha_i, 1 \leq i \leq n)$ be nonnegative real numbers. Let $s = \sum_1^n \alpha_i \chi(A_i) \in \mathcal{S}^+$ be the corresponding nonnegative simple function. We put

$$\int s d\mu = \sum_{i=1}^n \alpha_i \mu(A_i).$$

  Here, $\int s d\mu$ is called the *integral* of s with respect to μ. It can be checked that $\int s d\mu$ is independent of the decomposition of s and that $\int$ is linear and increasing on $\mathcal{S}^+$.

- If f is any nonnegative measurable function on $(X, \mathcal{A}, \mu)$, we put

$$\int f d\mu = \sup \int s d\mu, s \in \mathcal{S}^+, s \leq f.$$

- If f is any measurable function on $(X, \mathcal{A}, \mu)$, we define for every $x \in X$

$$\begin{aligned} f^+(x) &= \max(f(x), 0) \\ f^-(x) &= -\min(f(x), 0). \end{aligned}$$

  Then, f^+ and f^- are measurable and

$$f = f^+ - f^-.$$

  If $\int f^+ d\mu$ and $\int f^- d\mu$ are both finite, we say that f is *integrable* and we define

$$\int f d\mu = \int f^+ d\mu - \int f^- d\mu.$$

- If A is measurable and f is integrable, then $\chi_A f$ is integrable and we define

$$\int_A f d\mu = \int \chi_A f d\mu.$$

- If the reference to the measure is clear, we can simplify the notation by writing

$$\int f d\mu = \int f;$$

if the reference to the variable x in X is necessary, we write (using classical notations)

$$\int f d\mu = \int f(x) d\mu(x) = \int f(x) dx.$$

The main property of $\int$ is its linearity over $\mathbb{R}$. Moreover, it is an increasing function: if f and g are measurable functions on $(X, \mathcal{A}, \mu)$,

$$f \leq g \Rightarrow \int f d\mu \leq \int g d\mu.$$

The $\mathbb{R}$-vector space of integrable (real) functions on $(X, \mathcal{A}, \mu)$ is denoted by $\mathcal{L}(X, \mathcal{A}, \mu)$, or $\mathcal{L}(X)$ if no confusion is possible. The main tools of integration theory are convergence theorems. We cannot enumerate them all in our context (*Beppo Levi*'s theorem, *Fatou*'s lemma, etc.), and we only mention one of the most famous results, *Lebesgue's dominated convergence theorem*.

Theorem 15. *Let $(X, \mathcal{A}, \mu)$ be a measure space. Let g be a nonnegative integrable function and $(f_n)_{n \in \mathbb{N}}$ be a sequence of measurable functions on X such that*

$$\lim_{n \to \infty} f_n(x) = f(x)$$

for almost every $x \in X$. Suppose that, for almost every $x \in X$ and every $n \in N$,

$$|f_n(x)| \leq g(x).$$

Then, for every $n \in \mathbb{N}$, f_n is integrable, f is integrable, and

$$\lim_{n \to \infty} \int f_n d\mu = \int f d\mu.$$

5.3 The Standard *Lebesgue* Measure on $\mathbb{E}^N$

The purpose of this book being to assign to suitable subsets of $\mathbb{E}^N$ real numbers which measure their "shape" and satisfy such natural conditions as the inclusion–exclusion principle (see (1.2) in the introduction), consequence of an additivity property of an underlying measure, we need to use the classical measure in $\mathbb{E}^N$, i.e., the *Lebesgue* measure. Here, we would like to point out the difficulty of a coherent construction, due to the fact that one imposes the additivity. In fact, it is tempting to build a theory, in which one can assign a (non-null) "natural volume" to each subset of $\mathbb{E}^N$ which is both additive and invariant under rigid motion. However, it appears that this is impossible: for instance for $N = 3$, using the *axiom of choice*, one can prove that there does not exist any function

$$\mu : \mathcal{P}(\mathbb{E}^3) \to \mathbb{R}^{+*},$$

additive and invariant under rigid motion, and such that $\mu([0,1]^3) = 1$. This leads to the construction of the *Lebesgue measure* on $\mathbb{E}^N$.[2]

5.3.1 *Lebesgue Outer Measure on $\mathbb{R}$ and $\mathbb{E}^N$*

- We first define the outer measure

$$\mathcal{L}^* : \mathcal{P}(\mathbb{R}) \to \bar{\mathbb{R}}^+$$

on $\mathbb{R}$ as follows. If A is any subset of $\mathbb{R}$, $\mathcal{D}_A$ denotes the family of bounded open intervals $]a_i, b_i[$ such that $A \subset \cup_i^\infty]a_i, b_i[$. We put

$$\forall A \in \mathcal{P}(\mathbb{R}), \mathcal{L}^*(A) = \inf_{]a_i, b_i[\in \mathcal{D}(A)} \sum_i (b_i - a_i).$$

It is easy to check that $\mathcal{L}^*$ is an outer measure on $\mathbb{R}$ such that the measure of any interval is its usual length.

- This construction can be generalized to $\mathbb{E}^N$. We call a *parallelotope P* the product of N intervals $I_1 \times ... \times I_N$. We define the *$N$-volume* $\mathrm{Vol}_N(P)$ of P as the product $l_1...l_N$ of their lengths $l_i, 1 \le i \le N$:

$$\mathrm{Vol}_N(P) = l_1...l_N. \tag{5.8}$$

If A is any subset of $\mathbb{E}^N$, $\mathcal{D}_A$ denotes now the family of bounded open parallelotopes P_i such that $A \subset \cup_i^\infty P_i$. We put

$$\mathcal{L}^{N^*}(A) = \inf_{P_i \in \mathcal{D}_A} \sum_i \mathrm{Vol}_N(P_i).$$

The map $\mathcal{L}^{N^*}$ is an outer measure on $\mathbb{E}^N$ such that the measure of any parallelotope is its usual N-volume. This outer measure is called the *Lebesgue* outer measure on $\mathbb{E}^N$. Of course, one can identify $\mathbb{R}$ and $\mathbb{E}^1$, $\mathcal{L}^*$ and $\mathcal{L}^{1^*}$.

In some sense, there is only one outer measure with "good" properties on $\mathbb{E}^N$: it is the Lebesgue outer measure. This result can be formalized as follows.

Theorem 16. *The Lebesgue outer measure $\mathcal{L}^N$ is the unique Borel regular translation-invariant outer measure on $\mathbb{E}^N$, which assigns 1 to the unit N-cube.*

---

[2] We do not cover the extremely important Banach–Tarski paradox, beyond the scope of this book, see [80] for instance.

5.3.2 Lebesgue Measure on $\mathbb{R}$ and $\mathbb{E}^N$

Let $\mathcal{M}_{\mathcal{L}^{N*}}$ be the σ-algebra of all $\mathcal{L}^{N*}$-measurable sets. It can be checked that every Borel subset of $\mathbb{E}^N$ is $\mathcal{L}^{N*}$-measurable, i.e.,

$$\mathcal{B} \subset \mathcal{M}_{\mathcal{L}^{N*}},$$

where $\mathcal{B}$ denotes the collection of all Borel subsets of $\mathbb{E}^N$ (in other words, $\mathcal{L}^{N*}$ is Borel regular). The restriction of the outer measure $\mathcal{L}^{N*}$ to $\mathcal{M}_{\mathcal{L}^{N*}}$ (resp., to $\mathcal{B}$) is a measure which will be called the *Lebesgue measure* $\mathcal{L}^N$ on $(\mathbb{E}^N, \mathcal{M}_{\mathcal{L}^{N*}})$ (resp., $(\mathbb{E}^N, \mathcal{B})$).

A function f on $\mathbb{E}^N$ is called *Lebesgue measurable* (resp., *Borel measurable*), if the inverse image of any open subset of $\mathbb{R}$ is $\mathcal{L}^{N*}$-measurable (resp., a Borel subset of $\mathbb{E}^N$).

As for outer measure, the Lebesgue measure is the only one which has "good properties." This can be stated as follows.

Theorem 17. *The Lebesgue measure $\mathcal{L}^N$ is the only translation-invariant measure on $(\mathbb{E}^N, \mathcal{B})$, which assigns 1 to the unit N-cube.*

The Lebesgue measure on $(\mathbb{E}^N, \mathcal{B})$ is *regular* in the following sense:

- The Lebesgue measure of any compact subset of $\mathbb{E}^N$ is finite.
- For every measurable subset A, $\mathcal{L}^N(A) = \inf\{\mathcal{L}^N(U) : A \subset U, U \text{ open}\}$.
- For every open subset U, $\mathcal{L}^N(U) = \sup\{\mathcal{L}^N(K) : K \subset U, K \text{ compact}\}$.

Analogous results are true for the Lebesgue measure on $(\mathbb{E}^N, \mathcal{M}_{\mathcal{L}^{N*}})$.

Important Notation Usually, if A is any Borel subset of $\mathbb{E}^N$, $\mathcal{L}^N(A)$ is called the *N-volume* of A (or simply the *volume* of A if no confusion is possible), also denoted by $\mathrm{Vol}_N(A)$ (or simply $\mathrm{Vol}(A)$ if no confusion on the dimension is possible):

$$\mathcal{L}^N(A) = \mathrm{Vol}_N(A) = \mathrm{Vol}(A)$$

(if $N = 2$ it is the *area* of A and, if $N = 1$, the *length* of A).

5.3.3 Change of Variable

In this section, we give the classical theorem of change of variable in a general setting (see [43]). First of all, we need to define the *k-Jacobian $J_k f$* of a differentiable map $f : \mathbb{E}^n \to \mathbb{E}^p$:

- When $n = p$, this definition is the classical one.
- If f is linear, the k-Jacobian $J_k f$ of f is the maximal k-dimensional volume of the image of a k-dimensional cube.
- If f is differentiable at a, then $J_k f(a)$ is the k-Jacobian of $Df(a)$.

If no confusion is possible, we simplify the notation, writing

$$J_k f(a) = J_f.$$

The classical theorem of change of variable can be stated as follows. We set it in $(\mathbb{E}^n, \mathcal{B}, \mathcal{L}^n)$.

Theorem 18. *Let f be a differentiable one-to-one map of an open set U of $\mathbb{E}^n$ onto a bounded open set V of $\mathbb{E}^n$. If f^{-1} is continuous, then:*

1. For every Borel measurable function $g : \mathbb{E}^n \to \mathbb{R}$, the function

$$\phi : U \to \mathbb{E}^n$$

defined by

$$\phi(x) = g[f(x)]J(f)(x)$$

is Borel measurable.

2. Moreover,

$$\int_V g(y)d\mathcal{L}_y^n = \int_U g[f(x)]J(f)(x)d\mathcal{L}_x^n, \qquad (5.9)$$

where $\mathcal{L}_x^n$ (resp., $\mathcal{L}_y^n$) denotes the Lebesgue measure with respect to the variable x (resp., y).

5.4 Hausdorff Measures

There are much more general (outer) measures on $\mathbb{E}^N$. They are the *Hausdorff measures* $\mathcal{H}^k$, $1 \le k \le N$ [43]. Their definitions use the standard metric on $\mathbb{E}^N$ and the Lebesgue measure b_k of the unit ball of $\mathbb{E}^k$. They are intimately related to the notion of *dimension* of a subset of $\mathbb{E}^N$.

For each nonnegative number k, the k-Hausdorff measure $\mathcal{H}^k$ is defined as follows: if A is any subset of $\mathbb{E}^N$,

$$\mathcal{H}^k(A) = \lim_{\delta \to 0} \inf_{A \subset \cup P_j, \mathrm{diam}\,(P_j) \le \delta} \sum b_k \left(\frac{\mathrm{diam}\, P_j}{2}\right)^k.$$

The infimum is taken over all countable families (P_j) which cover A, and whose elements P_j have diameter less than δ (Fig. 5.1).

This definition is not so easy to understand. Let us emphasize on some remarkable properties:

- k may be any nonnegative *real number* (the measure is called *fractal* if k is not an integer).

Fig. 5.1 Here is a covering of a plane curve c by a finite set of (full) ellipses and discs of diameter less than δ. The length of c equals its Hausdorff measure $\mathcal{H}^1(c)$

- A nontrivial fact is that $\mathcal{H}^N$ and $\mathcal{L}^N$ coincide on the Borel σ-algebra $\mathcal{B}$:

$$\mathcal{H}^N(A) = \mathcal{L}^N(A) = \mathrm{Vol}_N(A),$$

  for any A belonging to the Borel σ-algebra $\mathcal{B}$.
- In particular, if γ is a (compact) smooth curve of $\mathbb{E}^N$, its length (in the sense of Sect. 2.1) coincides with its 1-Hausdorff measure. This property can be generalized to any submanifold of $\mathbb{E}^N$ (see Sect. 11.2).

5.5 Area and Coarea Formula

The formula of change of variable (5.9) can be improved by using both Lebesgue and Hausdorff measures. If A is any subset of $\mathbb{E}^n$ and y is any point of $\mathbb{E}^p$, we put

$$i(f_{|A}, y) = \mathrm{card}\,\{x \in A : f(x) = y\}.$$

Theorem 19. (The Area Formula) *Let* $f : \mathbb{E}^n \to \mathbb{E}^p\,(n \le p)$ *be a Lipschitz map:*

1. If A is any measurable subset of $\mathbb{E}^n$, then

$$\int_A J_n(f)(x)d\mathcal{L}^n x = \int_{\mathbb{E}^p} i(f_{|A}, y)d\mathcal{H}^n y. \tag{5.10}$$

2. If g is any integrable function on $\mathbb{E}^n$,

$$\int_{\mathbb{E}^n} g(x)J_n(f)(x)d\mathcal{L}^n x = \int_{\mathbb{E}^p} \sum_{x \in f^{-1}(y)} g(x)d\mathcal{H}^n y. \tag{5.11}$$

Theorem 20. (The Coarea Formula) *Let* $f : \mathbb{E}^n \to \mathbb{E}^p\,(n > p)$ *be a Lipschitz map. If A is any measurable subset of $\mathbb{E}^n$, then*

$$\int_A J_p(f)(x)d\mathcal{L}^n x = \int_{\mathbb{E}^p} \mathcal{H}^{n-p}(A \cap f^{-1}(y))d\mathcal{L}^p y. \tag{5.12}$$

5.6 Radon Measures

Using the topology of $\mathbb{E}^N$ (in particular, the fact that it is locally compact), one can specify particular signed measures with help of Borel subsets with compact closure.

Definition 16. A *Radon measure* denotes a countably additive function

$$\mu : \mathcal{CB} \to \mathbb{R},$$

where $\mathcal{CB}$ denotes the class of Borel subsets with compact closure.

In our context, the *curvature measures* to be introduced will be *Radon measures*.

5.7 Convergence of Measures

Since our goal in this book is to obtain convergence and approximation of *curvature measures*, we need to introduce the more general notion of convergence of measures. For simplicity, we only mention here the *weak convergence* of a sequence of finite (signed) measures defined on $(\mathbb{E}^N, \mathcal{B})$. Let us denote by $C_b(\mathbb{E}^N)$ the set of bounded continuous functions on $\mathbb{E}^N$ and by $C_c(\mathbb{E}^N)$ the set of continuous functions with compact support on $\mathbb{E}^N$.

Definition 17. Let $(\mu_n)_{n \in \mathbb{N}}$ be a sequence of (signed) measures (resp., *Radon measures*) defined on $(\mathbb{E}^N, \mathcal{B})$ and let μ be a (signed) measure (resp., *Radon measure*) defined on $(\mathbb{E}^N, \mathcal{B})$ (resp., $(\mathbb{E}^N, \mathcal{CB})$). The sequence $(\mu_n)_{n \in \mathbb{N}}$ weakly converges to μ if, for all functions $f \in C_c(\mathbb{E}^N)$,

$$\lim_{n \to \infty} \int_{\mathbb{E}^N} f \, d\mu_n = \int_{\mathbb{E}^N} f \, d\mu. \tag{5.13}$$

This convergence is called the *weak-* convergence* by some authors (see [4] for instance).

Note that if a sequence $(\mu_n)_{n \in \mathbb{N}}$ weakly converges to μ, this does not imply in general that, for every Borel subset B,

$$\lim_{n \to \infty} \mu_n(B) = \mu(B).$$

However, when $(\mu_n)_{n \in \mathbb{N}}$ and μ are measures in the sense of Sect. 5.1.2, and if the μ-measure of the (topological) boundary of B is null, one has the following classical result.

Theorem 21. *Let* $(\mu_n)_{n \in \mathbb{N}}$ *be a sequence of bounded measures[3] on* $(\mathbb{E}^N, \mathcal{B})$*, which converges to a bounded measure* μ *in the following sense: for all functions* $f \in C_b(\mathbb{E}^N)$,

---

[3] That is, for all $n \in \mathbb{N}$, $\mu_n : \mathcal{B} \to \mathbb{R}^+$ is a measure in the sense of Sect. 5.1.2.

$$\lim_{n \to \infty} \int_{\mathbb{E}^N} f d\mu_n = \int_{\mathbb{E}^N} f d\mu. \tag{5.14}$$

Then for every $B \in \mathcal{B}$ such that $\mu(\partial B) = 0$:

$$\lim_{n \to \infty} \mu_n(B) = \mu(B).$$

Unfortunately, Theorem 21 cannot be extended to the class of *signed* measures.

Part III
Background: Polyhedra and Convex Subsets

Chapter 6
Polyhedra

Although one could give a general abstract theory of polyhedra, we prefer to restrict our attention to subsets of $\mathbb{E}^N$.

6.1 Definitions and Properties of Polyhedra

We say that $(n+1)$ points of $\mathbb{E}^N$ are *geometrically independent* if they span an affine n-dimensional subspace of $\mathbb{E}^N$. Let $(v_0, v_1, ..., v_n)$ be $(n+1)$ *geometrically independent* points. The *n-simplex* spanned by $(v_0, v_1, ..., v_n)$ is the *convex hull* of $(v_0, v_1, ..., v_n)$, i.e., the set σ^n of all points $x \in \mathbb{E}^N$ which can be written as

$$x = \sum_{i=0}^{n} t_i v_i, \text{ where } \sum_{i=0}^{n} t_i = 1. \tag{6.1}$$

A simplex is obviously convex. Let us now recall standard definitions:

- The integer n is the *dimension* of σ^n.
- The points $v_0, v_1, ..., v_n$ are called the *vertices* of σ^n.
- A simplex spanned by a subset of $v_0, v_1, ..., v_n$ is a *face* of σ^n.
- The union of the faces of σ^n of dimension $k < n$ defines the *boundary* $\partial \sigma^n$ of σ^n.
- The interior $\text{int}(\sigma)$ of σ is the set $\sigma \backslash \partial \sigma$.
- Usually, 1-simplices σ^1 are called *edges* and denoted by the letter e.
- Usually, 2-simplices σ^2 are called *triangles* and denoted by the letter t.

Note that an n-simplex is homeomorphic to a (closed) ball $\mathbb{B}^n$, the homeomorphism sending the boundary of σ onto the boundary of $\mathbb{B}^n$ (Fig. 6.1).

We can now define the notion of *simplicial complex*.

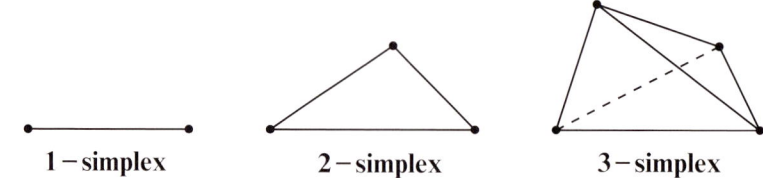

1 − simplex   **2 − simplex**   **3 − simplex**

Fig. 6.1 $1, 2, 3$-simplices in $\mathbb{E}^3$

Fig. 6.2 A two-dimensional simplicial complex

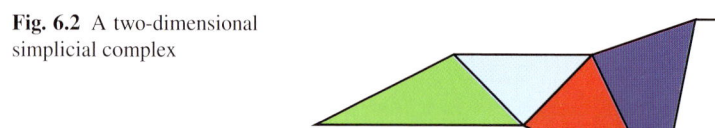

Definition 18. A simplicial complex C in $\mathbb{E}^N$ is a union of simplices of $\mathbb{E}^N$ such that:

- Every face of a simplex of C lies in C.
- The intersection of any two simplices of C is either empty or a face of each of them.

The dimension of C is the maximum of the dimensions of its simplices.

Note that a simplicial complex admits a natural topology.

This book will make use of different kinds of polyhedra (essentially polyhedra endowed with a triangulation). There are many different possible definitions (depending on the goal of the author). Let us give some general ones.

Definition 19.

- A triangulation of a topological space X is a pair (C, h), where C is a simplicial complex and h is an homeomorphism between C and X.
- A *polyhedron* embedded in $\mathbb{E}^N$ is a simplicial complex of $\mathbb{E}^N$ homeomorphic to a (piecewise linear) topological manifold (with or without boundary).

For instance, the simplicial complex of Fig. 6.2 is not a polyhedron, since it is not homeomorphic to a topological manifold.

Remark that (Fig. 6.3):

- By Definition 19, any polyhedron admits triangulations.
- A polyhedron in $\mathbb{E}^N$ admits a piecewise linear flat metric, since each of its faces is a portion of an affine (flat) subspace of $\mathbb{E}^N$.

We give the following definition.

Fig. 6.3 A two-dimensional polyhedron in $\mathbb{E}^3$ without boundary

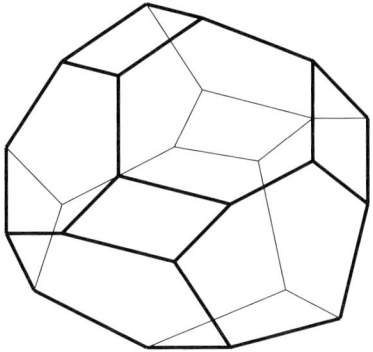

Fig. 6.4 The small black spherical triangle is the basis of the normal cone $L(e, \sigma^3)$ at q

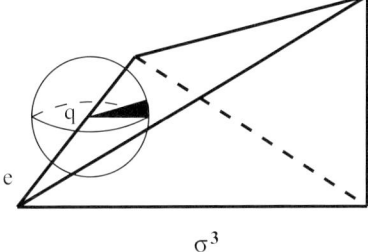

Definition 20. Let σ^l be an l-dimensional face of a k-simplex σ^k ($l < k$). Let $q \in \text{int}(\sigma^l)$. The following notions are independent of q:

1. The normal cone $C^\perp(\sigma^l, \sigma^k)$ to σ^l relative to the interior of σ^k is defined as follows:

$$C^\perp(\sigma^l, \sigma^k) = \{x \in \text{int}(\sigma^k) : \vec{q}\vec{x} \in \sigma^{l\perp}\}.$$

2. The basis of the normal cone $C^\perp(\sigma^l, \sigma^k)$ to σ^l is

$$L(\sigma^l, \sigma^k) = C^\perp(\sigma^l, \sigma^k) \cap \mathbb{S}^{k-l-1},$$

   where $\mathbb{S}^{k-l-1}$ is the unit sphere centered at q (Fig. 6.4).

3. The internal dihedral angle $\widetilde{(\sigma^l, \sigma^k)}$ is the measure of $L(\sigma^l, \sigma^k)$ with respect to the standard measure on $\mathbb{S}^{k-l-1}$.

4. The normalized internal dihedral angle (σ^l, σ^k) is the quotient of $\widetilde{(\sigma^l, \sigma^k)}$ by the standard measure s_{k-l-1} of $\mathbb{S}^{k-l-1}$:

$$(\sigma^l, \sigma^k) = \frac{\widetilde{(\sigma^l, \sigma^k)}}{s_{k-l-1}}.$$

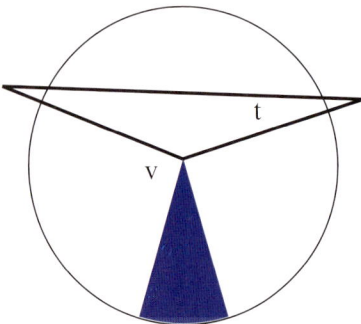

5. The external dihedral angle $\widetilde{(\sigma^l, \sigma^k)}^*$ is the measure of the subset of $\mathbb{S}^{k-l-1}$ obtained by intersecting $\mathbb{S}^{k-l-1}$ with the half-lines whose origin is q, and making an angle greater than $\frac{\pi}{2}$ with the interior of σ^k (Fig. 6.5).

6. The normalized external dihedral angle $(\sigma^l, \sigma^k)^*$ is the quotient of $\widetilde{(\sigma^l, \sigma^k)}^*$ by the standard measure s_{k-l-1} of $\mathbb{S}^{k-l-1}$:

$$(\sigma^l, \sigma^k)^* = \frac{\widetilde{(\sigma^l, \sigma^k)}^*}{s_{k-l-1}}.$$

This definition generalizes directly the angular defect for vertices of one- or two-dimensional polyhedra of $\mathbb{E}^3$, mentioned in Sect. 3.2.3.

6.2 Euler Characteristic

If C is a simplicial complex, the *Euler characteristic* $\chi(C)$ of C is the integer

$$\chi(C) = \sum_i (-1)^i \alpha_i(C), \tag{6.2}$$

where $\alpha_i(C)$ is the number of i-simplices in C. A simple computation shows that $\chi(C)$ is independent of the decomposition of C into simplices (Fig. 6.6).

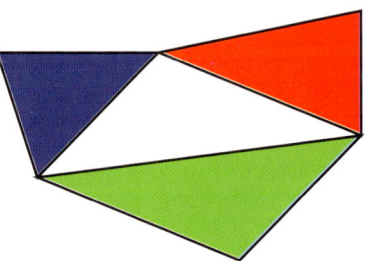

Fig. 6.6 The Euler characteristic of this gray triangulation C is $\chi(C) = 6 - 9 + 3 = 0$

6.3 Gauss Curvature of a Polyhedron

We introduce now the discrete notion of *Gauss curvature* of a polyhedron at each vertex σ^0, generalizing Definition 7. We shall see in the following chapters the relations between this discrete notion and the classical Gauss (or total) curvature of a smooth submanifold at a point.

Definition 21. If v is a vertex of a simplicial complex C, the Gauss curvature of v is the real number

$$G_v = \sum_i (-1)^i \sum_{v \in \sigma^i \subset C} \widetilde{(v, \sigma^i)}^*. \tag{6.3}$$

Let us now show that this point of view is coherent with Sect. 3.4. Suppose that P is a two-dimensional polyhedron of $\mathbb{E}^3$ and v is a vertex of P. Let $t_i, 1 \leq i \leq p$, be the triangles incident to v and α_i be the angle of t_i at v. Then[1]

$$G_v = 1 - \frac{p}{2} + \frac{1}{2\pi} \sum_{i=1}^{p} (\pi - \alpha_i),$$

i.e. (Fig. 6.7),

$$G_v = \frac{1}{2\pi} [2\pi - \sum_{i=1}^{p} \alpha_i]. \tag{6.4}$$

A discrete version of the Gauss–Bonnet theorem relates the Euler characteristic of a simplicial complex C to the (normalized) exterior angles of the vertices of C [8].

Theorem 22. (Gauss–Bonnet Theorem) *Let C be any simplicial complex (Fig. 6.8). Then,*

$$\sum_{v \in C} G_v = \chi(C). \tag{6.5}$$

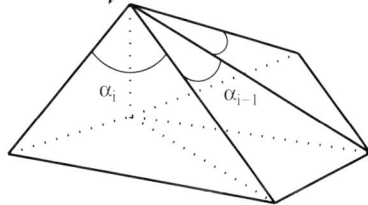

Fig. 6.7 By definition, the discrete Gauss curvature G_v of this two-dimensional polyhedron at the vertex v is $\frac{1}{2\pi}[2\pi - \sum_{i=1}^{p} \alpha_i]$

---

[1] This computation uses the values of the volumes of $\mathbb{S}^0$ and $\mathbb{S}^1$, given in Chap. 10. In particular, the volume of S^0 is 2, since $\mathbb{S}^0$ is reduced to two points.

Fig. 6.8 This is a reconstruction of a heart obtained by the triangulation of a cloud of points. This image is courtesy of Herve Delingette, I.N.R.I.A. Asclepios Team

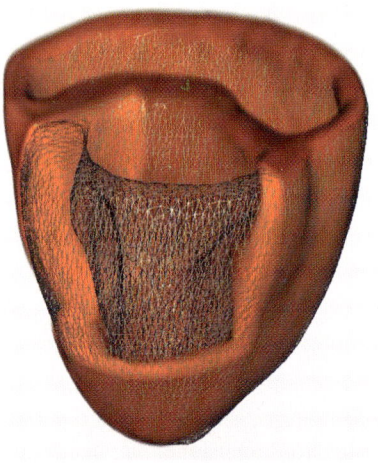

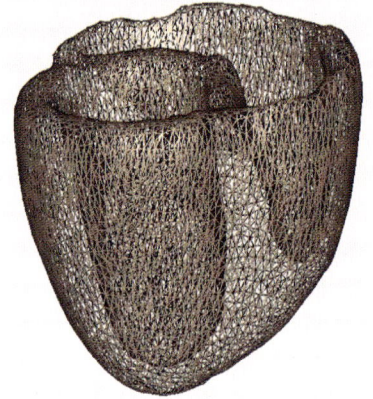

Chapter 7
Convex Subsets

There is an abundant literature on convexity, crucial in many fields of mathematics. We shall mention the basic definitions and some fundamental results (without proof), useful for our topic. In particular, we shall focus on the properties of the volume of a convex body and its boundary. The reader can consult [9, 71, 74, 79] for details.

7.1 Convex Subsets

7.1.1 Definition and Basic Properties

Definition 22.

- A subset $\mathcal{K}$ of $\mathbb{E}^N$ is called convex if for all p and q in $\mathcal{K}$, the segment $[p, q]$ lies in $\mathcal{K}$ (Fig. 7.1).
- A compact convex subset $\mathcal{K}$ of $\mathbb{E}^N$ is called a convex body if it has a nonempty interior (Fig. 7.2).
- The boundary $\partial \mathcal{K}$ of a convex body $\mathcal{K}$ is called a convex hypersurface.

We shall denote by $\mathcal{C}$ (or $\mathcal{C}_{\mathbb{E}^N}$ if we need to be more precise) the space of convex subsets of $\mathbb{E}^N$. We denote by $\mathcal{C}_b$ (resp., $\mathcal{C}_c$) the space of convex bodies (resp., compact convex subsets) of $\mathbb{E}^N$. The space $\mathcal{C}_c$, endowed with the Hausdorff metric, is a complete metric space [9, Tome 3, 12.9.1.2].

Proposition 1. *Let $(\mathcal{K}_n)_{n \in \mathbb{N}}$ be a sequence of convex subsets of $\mathbb{E}^N$ whose Hausdorff limit is the subset $\mathcal{K}$. Then, $\mathcal{K}$ is convex.*

Moreover, any convex body of $\mathbb{E}^N$ can be approximated by compact convex polyhedra [9, Tome 3, 12.9.2.1].

Proposition 2. *Let $\mathcal{K}$ be a convex body of $\mathbb{E}^N$. Then, $\mathcal{K}$ is the Hausdorff limit of a sequence of compact convex polyhedra (Fig. 7.3).*

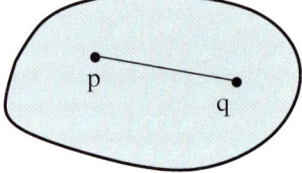 

Fig. 7.1 The subset on the *left* is a convex subset of $\mathbb{E}^2$, the one on the *right* is nonconvex

Fig. 7.2 A convex body $\mathcal{K}$ in $\mathbb{E}^3$

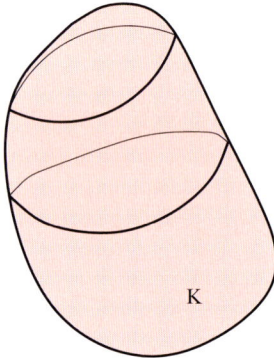

Fig. 7.3 A sequence of convex subsets $\mathcal{K}_n$ in $\mathbb{E}^2$ tending to the (convex) ellipse E

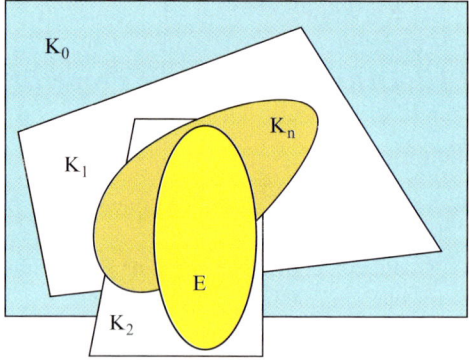

For our purpose, the orthogonal projection (with respect to the usual scalar product of $\mathbb{E}^N$) is one of the main tools. We mention here classical properties of this map in the convex context. Basically, we need a deep study of two types of projection:

1. The first one is the projection of a convex subset onto an (affine) subspace of $\mathbb{E}^N$ [9, Tome 2, 9.11.6 and Tome 3].

   Proposition 3. *Let P be any vector (or affine) subspace of $\mathbb{E}^N$ and let $\mathcal{K}$ be a convex subset of $\mathbb{E}^N$. Then:*

   - *The orthogonal projection $\mathrm{pr}(\mathcal{K})$ of $\mathcal{K}$ onto P is convex (Fig. 7.4).*
   - *The map $\mathrm{pr} : \mathcal{C}_{\mathbb{E}^N} \to \mathcal{C}_P$ is continuous for the Hausdorff topologies.*

Fig. 7.4 The projection of a
convex subset onto an affine
subspace is convex

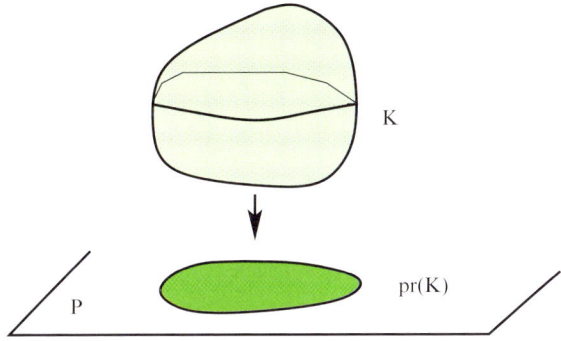

Fig. 7.5 The point p has a
unique orthogonal projection
onto the convex subset $\mathcal{K}$

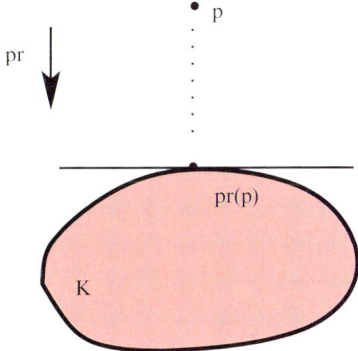

2. The second one is the projection of $\mathbb{E}^N$ onto a convex subset.

Proposition 4. *Let $\mathcal{K}$ be a compact convex subset of $\mathbb{E}^N$. Then every point of $\mathbb{E}^N$ has a unique orthogonal projection onto $\mathcal{K}$. This map is continuous on $\mathbb{E}^N$ and differentiable on $\mathbb{E}^N \setminus \mathcal{K}$ (Fig. 7.5).*

For details of the proofs, see [9].

7.1.2 The Support Function

Let us now introduce the *support function* of a convex body $\mathcal{K}$ of $\mathbb{E}^N$ (see [71, Sect. 13] for instance). We denote by $h_{\mathcal{K}}$ the *support function* of $\mathcal{K}$, i.e., the map

$$h_{\mathcal{K}} : \mathbb{S}^{N-1} \to \mathbb{R},$$

defined by

$$h_{\mathcal{K}}(\xi) = \max_{m \in \mathcal{K}} < m, \xi > .$$

Fig. 7.6 The support function. Here, $h_K(\xi) = <m_0, \xi>$

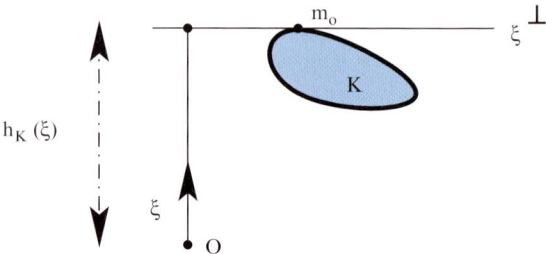

For every $\xi \in \mathbb{S}^{N-1}$, $\mathcal{K}$ lies in one of the two half-spaces determined by the *support hyperplane* (Fig. 7.6)

$$H = \{m \in \mathbb{E}^N : <m, \xi> = h_{\mathcal{K}}(\xi)\}.$$

A convex body is completely determined by its support function (see [9, Tome 3, 11.5, p. 43] or [74, p. 38]). Moreover, if $\mathcal{K}$ and $\mathcal{L}$ are two convex bodies, then their Hausdorff distance $d(\mathcal{K}, L)$ can be evaluated in terms of their support functions:

$$d(\mathcal{K}, \mathcal{L}) = \sup_{\xi \in \mathbb{S}^{N-1}} |h_{\mathcal{K}}(\xi) - h_{\mathcal{L}}(\xi)|.$$

This implies the following proposition.

Proposition 5. *The topology determined by the Hausdorff distance coincides with the topology of uniform convergence of support functions on the class of (compact) convex subsets of* $\mathbb{E}^N$.

7.1.3 The Volume of Convex Bodies

We shall concentrate on the restriction of the N-volume functional Vol_N to the class $\mathcal{C}_c$ of compact convex subsets of $\mathbb{E}^N$. As we have seen in Chap. 5, the volume defined for subsets of $\mathbb{E}^N$ can be considered without distinction as the Lebesgue measure or as the N-Hausdorff measure on $\mathbb{E}^N$.

The N-volume functional is obviously not continuous on the class of compact subsets of $\mathbb{E}^N$ endowed with the Hausdorff topology. A simple argument is to say that, if X is a compact subset of $\mathbb{E}^N$, the finite subsets of X are dense in the set of compact subsets of X, endowed with the Hausdorff topology, and the N-volume of each finite subset is null (see [9, Tome 2, Chap. 9] for instance). However, the restriction of the N-volume to $\mathcal{C}_c$ is continuous for the Hausdorff topology.

Theorem 23. *If $\mathcal{K}_n$ is a sequence of (compact) convex subsets whose Hausdorff limit is the convex subset $\mathcal{K}$, then*

$$\lim_{n \to \infty} \text{Vol}_N(\mathcal{K}_n) = \text{Vol}_N(\mathcal{K}). \tag{7.1}$$

Fig. 7.7 A convex subset K
between two close polyhedra

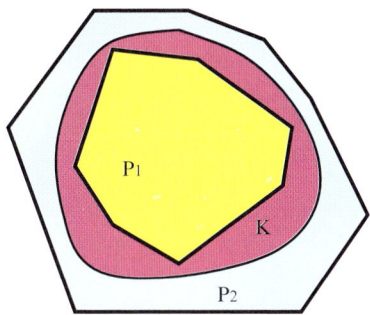

Sketch of proof of Theorem 23 We apply the following result (whose proof is easy and left to the reader). □

Lemma 2. *If K is a compact convex subset of $\mathbb{E}^N$ and ε is a positive number, there exist two polyhedra P_1, P_2 such that $P_1 \subset K \subset P_2$ and*

$$\partial P_1 \cap K = \partial P_2 \cap K = \emptyset,$$

satisfying (Fig. 7.7)

$$|\mathrm{Vol}_N(P_1) - \mathrm{Vol}_N(P_2)| \leq \varepsilon. \tag{7.2}$$

Now, let K_n be a sequence of compact convex subsets of $\mathbb{E}^N$ whose Hausdorff limit is a compact convex subset K. Let $\varepsilon > 0$ with P_1, P_2 as before. If we choose p large enough to ensure that $d(K, K_p) \leq \eta$ and $P_1 \subset K_p \subset P_2$, then

$$|\mathrm{Vol}_N(K_p) - \mathrm{Vol}_N(K)| \leq |\mathrm{Vol}_N(P_1) - \mathrm{Vol}_N(P_2)| \leq \varepsilon. \tag{7.3}$$

Then

$$\lim_{n \to \infty} \mathrm{Vol}_N(K_n) = \mathrm{Vol}_N(K).$$

For details, the reader can consult [9, Tome 3, 12.9.3].

7.2 Differential Properties of the Boundary

We now deal with the (topological) boundary ∂K of a convex body of K of $\mathbb{E}^N$. By definition, ∂K is the subset of points belonging to the closure of K and to the closure of the complement of K. Locally, ∂K is the graph of a Lipschitz function defined over one of its support hyperplanes H. It can be proved that this function is (convex and) twice differentiable $\mathcal{L}^{N-1}$-almost everywhere on H (see [1, 72]). We deduce the following theorem.

Theorem 24. *The boundary ∂K of a convex body K of $\mathbb{E}^N$ is $\mathcal{L}^{N-1}$-almost everywhere a twice differentiable hypersurface.*

Consequently, the boundary $\partial \mathcal{K}$ of a convex body $\mathcal{K}$ admits a unique outward unit normal vector field ξ defined on an open subset $U_{\partial \mathcal{K}}$ of $\partial \mathcal{K}$ of full measure,[1] on which it is continuous. Then, at each point m of $U_{\partial \mathcal{K}}$, one can construct the outward unit normal vector ξ_m and the normal line L_m to $\partial \mathcal{K}$ at m.

We mention (without proof) the classical theorem.

Theorem 25. *Let $\mathcal{K}_n$ be a sequence of convex bodies of $\mathbb{E}^N$ whose Hausdorff limit is the convex body $\mathcal{K}$. Let $m \in U_{\partial K}$ and let m_n be the sequence of points which are the intersection of L_m with $\partial \mathcal{K}_n$. If $m \in U_{\partial K}$ (resp., $m_n \in U_{\partial \mathcal{K}_n}$ for every $n \in \mathbb{N}$), then (with obvious notations)*

$$\lim_{n \to \infty} \xi_{m_n} = \xi_m.$$

7.3 The Volume of the Boundary of a Convex Body

If $\mathcal{K}$ is a convex body, Sect. 7.2 shows that there are many ways to compute the $(N-1)$-volume of $\partial \mathcal{K}$:

1. We can compute its $(N-1)$-Hausdorff measure $\mathcal{H}^{N-1}(\partial \mathcal{K})$, considering $\partial \mathcal{K}$ as any subset of $\mathbb{E}^N$.
2. We can avoid the points of $\partial \mathcal{K}$ which are not differentiable and compute the $(N-1)$-volume of the resulting (smooth) hypersurface.
3. We can compute the supremum of the $(N-1)$-volume of the boundary of the compact convex polyhedral bodies included in $\mathcal{K}$.
4. We can compute the infimum of the $(N-1)$-volume of the boundary of the compact convex polyhedral bodies which contain $\mathcal{K}$.

It appears that *all these four quantities coincide*. It is the $(N-1)$-volume $\mathrm{Vol}_{N-1}(\partial \mathcal{K})$ of $\partial \mathcal{K}$. This result is not trivial, beyond the scope of this book [9, Tome 2 and 3].

Let us focus our attention on two properties of the boundary of a convex body:

1. The first asserts that one can approximate the $(N-1)$-volume of the boundary $\partial \mathcal{K}$ of a convex body $\mathcal{K}$ by the $(N-1)$-volume of the boundary of another convex body "close to it."
2. The second asserts that one obtains the $(N-1)$-volume of the boundary $\partial \mathcal{K}$ of a convex body $\mathcal{K}$ by integration over all hyperplanes $\mathcal{H}$ of the $(N-1)$-volume of the projection of $\mathcal{K}$ onto $\mathcal{H}$.

Let us now be more precise:

1. **A convergence result**

   Theorem 26. *Let $\mathcal{K}_n$ be a sequence of convex bodies of $\mathbb{E}^N$ whose Hausdorff limit is the convex body $\mathcal{K}$. Then*

---

[1] That is, such that the $(N-1)$-volume of $U_{\partial \mathcal{K}}$ equals the $(N-1)$-volume of $\partial \mathcal{K}$.

$$\lim_{n\to\infty} \text{Vol}_{N-1}(\partial\mathcal{K}_n) = \text{Vol}_{N-1}(\partial\mathcal{K}).$$

Note that this result is wrong if the $\mathcal{K}_n$ are not convex (see the example of the Lantern of Schwarz in Sect. 3.1.3).

Sketch of proof of Theorem 26 Using the *area formula* (see Theorem 19), this result can be deduced from Theorem 25: consider the orthogonal projection pr_n of $\mathcal{K}_n$ onto $\mathcal{K}$. Its Jacobian J_n tends to 1 when n tends to infinity. Consequently,

$$\lim_{n\to\infty} \text{Vol}_{N-1}(\partial\mathcal{K}_n) = \lim_{n\to\infty} \int_{\mathcal{K}_n} dv_{\mathcal{K}_n} =$$

$$\lim_{n\to\infty} \int_{\mathcal{K}} J_n dv_{\mathcal{K}} = \int_{\mathcal{K}} dv_{\mathcal{K}} = \text{Vol}_{N-1}(\partial\mathcal{K}).$$

Another direct way is to use the definition of the $(N-1)$-volume in terms of supremum (or infimum) of polyhedra included in (or containing) $\mathcal{K}$. See also [9, Tome 3, 12.10]. $\square$

2. **An integral formula**

The *Cauchy formula* relates the $(N-1)$-volume of $\partial\mathcal{K}$ and the $(N-1)$-volume of the projections of $\mathcal{K}$ onto hyperplanes of $\mathbb{E}^N$ (see [73, Chap. 13] for details). The formula can be stated as follows.

Theorem 27. (Cauchy Formula) *Let $\mathcal{K}$ be a convex body of $\mathbb{E}^N$. Then*

$$\text{Vol}_{N-1}(\partial\mathcal{K}) = \frac{1}{(N-1)s_{N-1}} \int_{\xi\in\mathbb{S}^{N-1}} \text{Vol}_{N-1}(\mathcal{K}_{p\xi^\perp})d\xi, \qquad (7.4)$$

where $\mathcal{K}_{p\xi^\perp}$ denotes the orthogonal projection of $\mathcal{K}$ onto the hyperplane $\xi^\perp$.

Of course, using the twofold covering of $\mathbb{S}^{N-1}$ onto the Grassmann manifold $G(N,N-1)$ of (unoriented) $(N-1)$-subspaces of $\mathbb{E}^N$, one can integrate over $G(N,N-1)$, endowed with its standard invariant Haar measure (see Chap. 10). Since the volume of $G(N,N-1)$ is one half the volume of $\mathbb{S}^{N-1}$, one obtains

$$\text{Vol}_N(\partial\mathcal{K}) = \frac{2}{(N-1)s_{N-1}} \int_{G(N,N-1)} \text{Vol}_{N-1}(\mathcal{K}_{P^{N-1}})dv_{G(N,N-1)}, \qquad (7.5)$$

where $\mathcal{K}_{p^{N-1}}$ denotes the projection of $\mathcal{K}$ onto the hyperplane P^{N-1}.

Sketch of proof of Theorem 27

- This theorem is easy to prove for a convex polyhedron $\mathcal{K}$:

  (a) In fact, almost every $(N-1)$-hyperplane P of $\mathbb{E}^N$ is transversal to the boundary of $\mathcal{K}$. The sum of the volumes of the projections of the $(N-1)$-faces f of the boundary of $\mathcal{K}$ onto P equals twice the volume of the projection of $\mathcal{K}$ onto P. Then we have

$$\sum_f \text{Vol}_{N-1}(\text{pr}_P f) = 2\text{Vol}_{N-1}(\text{pr}_P \mathcal{K}). \qquad (7.6)$$

(b) Note that, for every $(N-1)$-face f of $\partial\mathcal{K}$, we have

$$\mathrm{Vol}_{N-1}(\mathrm{pr}_P f) = |<\xi,\nu>|\mathrm{Vol}_{N-1}(f), \qquad (7.7)$$

where ξ is a unit vector normal to P and ν is a unit vector normal to f. Then

$$\int_{G(N,N-1)} \mathrm{Vol}_{N-1}(\mathcal{K}_{P^{N-1}}) dv_{G(N,N-1)}$$

$$= \frac{1}{2}\sum_f \mathrm{Vol}_{N-1}(f) \int_{\xi\in\mathbb{S}^{N-1}} |<\xi,n_f>| dv_{\mathbb{S}^{N-1}}.$$

(c) We conclude by using the twofold covering of $G(N,N-1)$ onto $\mathbb{S}^{N-1}$ and using the well-known following equality: for any fixed unit vector ν,

$$\int_{\xi\in\mathbb{S}^{N-1}} |<\xi,\nu>| dv_{\mathbb{S}^{N-1}} = 2(N-1)s_{N-1}, \qquad (7.8)$$

where s_{N-1} denotes the $(N-1)$-volume of $\mathbb{S}^{N-1}$.

• We extend this formula to any convex body by continuity of the $(N-1)$-volume function over the convex subsets of $\mathbb{E}^N$, using Theorem 26: let $\mathcal{K}$ be a convex body of $\mathbb{E}^N$ and let $(\mathcal{K}_n)_{n\in\mathbb{N}}$ be a sequence of convex bodies with polyhedral boundaries whose Hausdorff limit is $\mathcal{K}$. Then, for every hyperplane P^{N-1} of $\mathbb{E}^N$,

$$\lim_{n\to\infty} \mathrm{Vol}_{N-1}(\mathcal{K}_{n_{P^{N-1}}}) = \mathrm{Vol}_{N-1}(\mathcal{K}_{P^{N-1}}).$$

On the other hand,

$$\lim_{n\to\infty} \mathrm{Vol}_{N-1}(\partial\mathcal{K}_n) = \mathrm{Vol}_{N-1}(\partial\mathcal{K}),$$

from which we deduce Theorem 27 in the general case. $\qquad\square$

7.4 The Transversal Integral and the Hadwiger Theorem

A general introduction to *valuations* can be found in [56] and an introduction to the integral geometry of convex sets can be found in [73, Chap. 13] with all details.[2]

7.4.1 Notion of Valuation

We restrict our attention to the set $\mathcal{C}_c$ of compact convex subsets of $\mathbb{E}^N$.

---

[2] The indexing conventions are different in these two books. This may induce confusion for the reader.

Definition 23. A valuation on $\mathcal{C}_c$ is a map

$$\Phi : \mathcal{C}_c \to \mathbb{R},$$

such that:

- For all $A, B \in \mathcal{C}$ such that $A \cup B$ is convex,

$$\Phi(A \cup B) = \Phi(A) + \Phi(B) - \Phi(A \cap B).$$

- $\Phi(\emptyset) = 0$.

Note that the N-volume defines a valuation on $\mathcal{C}_c$ as well as the $(N-1)$-volume of the boundary. We have seen that the $(N-1)$-volume of the boundary of a compact convex subset of $\mathbb{E}^N$ can be expressed via the *Cauchy formula* (Theorem 27) as an integral on a suitable Grassmannian manifold. We shall see in Sect. 7.4.2 that all the valuations on $\mathcal{C}_c$ can be obtained in such a way.

7.4.2 Transversal Integral

We introduce particular valuations called the *transversal integral*. They are the integral of the volume of the projections of a convex body over $G(N, k)$ (where $G(N, k)$ denotes the Grassmannian of (unoriented) k-planes in $\mathbb{E}^N$, endowed with its standard Haar measure).

Definition 24. Let $\mathcal{K}$ be a convex subset of $\mathbb{E}^N$. Let P^{N-k} be any $(N-k)$-vector subspace of $\mathbb{E}^N$. Let $\mathcal{K}_{P^{N-k}}$ be the orthogonal projection of $\mathcal{K}$ onto P^{N-k}. The kth-transversal integral of $\mathcal{K}$ $(1 \leq k \leq N)$ is the real number

$$\mathbb{W}_k(\mathcal{K}) = c(N, k) \int_{G(N, N-k)} \mathrm{Vol}_{N-k}(\mathcal{K}_{P^{N-k}}) dv_{G(N, N-k)}, \tag{7.9}$$

with

$$c(N, k) = \frac{(N-k)}{N} \frac{s_k \ldots s_0}{s_{n-2} \ldots s_{N-k-1}}$$

(s_i denoting the i-volume of the unit i-dimensional sphere).

By convention, we put $\mathbb{W}_0(\mathcal{K}) = \mathrm{Vol}(\mathcal{K})$. Note that $N\mathbb{W}_1(\mathcal{K})$ is the $(N-1)$-volume of the boundary of $\mathcal{K}$. Moreover, the $\mathbb{W}_k$ satisfy the following properties.

Proposition 6. *For every* $k, 0 \leq k \leq N$:

- $\mathbb{W}_k$ *is a valuation on* $\mathcal{C}_c$.
- $\mathbb{W}_k$ *is invariant under rigid motions.*

- $\mathbb{W}_k$ is continuous on $\mathcal{C}_c$ for the Hausdorff topology.
- $\mathbb{W}_k$ is homogeneous of degree $N - k$, i.e.,

$$\forall \alpha \in \mathbb{R}, \mathbb{W}_k(\alpha \mathcal{K}) = \alpha^{N-k} \mathbb{W}_k(\mathcal{K}).$$

The proof is easy and left to the reader. However, we need to insist on the fact that the continuity property is a consequence of the continuity of the volume in $\mathcal{C}_c$ for the Hausdorff topology.

7.4.3 The Hadwiger Theorem

The crucial point is now the following result [55, 9.1.1].

Theorem 28. of Hadwiger *The valuations* $(\mathbb{W}_k)_{0 \le k \le N}$ *(defined on the class* $\mathcal{C}_c$ *of compact convex subsets of* $\mathbb{E}^N$*) form a basis of the vector space of all valuations on* $\mathcal{C}_c$*, continuous for the Hausdorff topology and invariant under rigid motions. Moreover, if* Φ *is any valuation on* $\mathcal{C}_c$ *continuous for the Hausdorff topology, invariant under rigid motions, and homogeneous of degree* $N - k$*, then there exists a real number* c *such that* $\Phi = c\mathbb{W}_k$*.*

The proof of Theorem 28 is a consequence of the following result.

Proposition 7. *Let* Φ *be any continuous valuation invariant under rigid motions on* $\mathcal{C}_c$*. If* Φ *is null on all (compact) convex bodies with empty interior, then* Φ *is proportional to the volume, i.e., there exists a constant* λ_N *such that for every (compact) convex body* $\mathcal{K}$*,*

$$\Phi(\mathcal{K}) = \lambda_N \text{Vol}(\mathcal{K}).$$

Sketch of proof of Proposition 7 The proof is long and technical. Let us only give an idea. Consider the unit cube $[0,1]^N$. Put

$$\Phi([0,1]^N) = \lambda_N.$$

Using the additivity of the valuation Φ, we see easily that

$$\Phi([0,1/k]^N) = k^{-N}\lambda_N.$$

It remains to deduce that

$$\Phi(B) = \lambda_N \text{vol}(B)$$

for any box B and then that

$$\Phi(\mathcal{K}) = \lambda_N \text{vol}(\mathcal{K})$$

for any convex body. The difficulty comes from the fact that convexity is not stable by union. The reader can consult [56, Theorem 8.1.4 and Chap. 9] for details. □

Using Proposition 7, Theorem 28 can be proved by induction: let Φ be any continuous valuation invariant under rigid motions on the set of convex bodies of $\mathbb{E}^N$. Let P be any hyperplane of $\mathbb{E}^N$ and denote by $\Phi_{|P}$ the restriction of Φ to P. If C is any convex body of P, then by assumption

$$\Phi_{|P}(C) = \sum_{i=1}^{N-1} \lambda_i \Phi_i(C).$$

Consider now the map

$$\Phi - \sum_{i=1}^{N-1} \lambda_i \Phi_i$$

as a valuation invariant under rigid motions on the set of all convex subsets of $\mathbb{E}^N$. It is clear that the (continuous) valuation $\Phi - \sum_{i=1}^{N-1} \lambda_i \Phi_i$ is null for any convex subset of $\mathbb{E}^N$ which is not a body. Since $\Phi - \sum_{i=1}^{N-1} \lambda_i \Phi_i$ is continuous, we deduce that $(\Phi - \sum_{i=1}^{N-1} \lambda_i \Phi_i)$ is proportional to the volume Φ_N of $\mathbb{E}^N$:

$$\Phi - \sum_{i=1}^{N-1} \lambda_i \Phi_i = \lambda_N \Phi_N.$$

This implies that

$$\Phi = \sum_{i=1}^{N} \lambda_i \Phi_i.$$

Remark It is interesting to note that Klain [55] proved in 1995 that Theorem 28 and Proposition 7 are actually equivalent.

As an application of the *Hadwiger* theorem, we can deduce the *Kubota* formula (see [73, p. 217] or [55, p. 126]).

Theorem 29. (Kubota Formula) *Let $\mathcal{K}$ be a convex body of $\mathbb{E}^N$. Then*

$$\mathbb{W}_k(\mathcal{K}) = \frac{2(N-1)}{ns_{N-2}} \int_{G(N,N-1)} \mathbb{W}_{k-1}(\mathcal{K}_{P^{N-1}}) dv_{G(N,N-1)}. \tag{7.10}$$

Noting that $\mathbb{S}^{N-1}$ is a twofold covering of $G(N, N-1)$, Theorem 29 can be stated as follows:

$$\mathbb{W}_k(\mathcal{K}) = \frac{(N-1)}{ns_{N-2}} \int_{\xi \in \mathbb{S}^{N-1}} \mathbb{W}_{k-1}(\mathcal{K}_{\xi^{\perp}}) d\xi. \tag{7.11}$$

Proof of Theorem 29 The map

$$\mathcal{V} : \mathcal{C}_c \to \mathbb{R}$$

defined by

$$\mathcal{V}(\mathcal{K}) = \int_{G(N,N-1)} \mathbb{W}_{k-1}(\mathcal{K}_{P^{N-1}}) dP^{N-1}$$

is continuous (for the Hausdorff topology), invariant under rigid motions, and homogeneous of degree $N - k$. Then, we can apply Theorem 28, which shows that $\mathcal{V}$ is proportional to $\mathbb{W}_k$. The constant of proportionality can be computed explicitly by testing the formula on a ball. $\square$

Part IV
Background: Classical Tools in Differential Geometry

Chapter 8
Differential Forms and Densities on $\mathbb{E}^N$

Curvature measures will be defined by integrating *differential forms*. Let us introduce their definitions, beginning with exterior algebra in a vector space and continuing with the smooth category. We only give here a brief survey. See [59] for a complete one.

8.1 Differential Forms and Their Integrals

8.1.1 Differential Forms on $\mathbb{E}^N$

1. **The algebraic point of view.** Let $\mathbb{E}^N$ be the N-dimensional Euclidean space and $\mathbb{E}^{N^*}$ be its dual. We denote by $\Lambda\mathbb{E}^N$ the exterior algebra of $\mathbb{E}^N$ and by $\Lambda^*\mathbb{E}^N$ the exterior algebra of $\mathbb{E}^{N^*}$. As usual, one can write

$$\Lambda\mathbb{E}^N = \oplus_{k=0}^N \Lambda_k\mathbb{E}^N,$$

where $\Lambda_0\mathbb{E}^N = \mathbb{R}$ and $\Lambda_k\mathbb{E}^N$ denotes the C_N^k-dimensional vector space spanned by elements of the form $v_1 \wedge \ldots \wedge v_k$, each $v_i(1 \leq i \leq k)$ lying in $\mathbb{E}^N$.

Similarly, we denote by $\Lambda^*\mathbb{E}^N$ the dual space of $\Lambda\mathbb{E}^N$ and, for all k, by $\Lambda^k\mathbb{E}^N$ the dual space of $\Lambda_k\mathbb{E}^N$. We have

$$\Lambda^*\mathbb{E}^N = \oplus_{k=0}^N \Lambda^k\mathbb{E}^N,$$

where $\Lambda^0\mathbb{E}^N = \mathbb{R}$ and $\Lambda^k\mathbb{E}^N$ denotes the C_N^k-dimensional vector space spanned by elements of the form $v_1^* \wedge \ldots \wedge v_k^*$, each $v_i^*(1 \leq i \leq k)$ lying in $\mathbb{E}^{N^*}$.

A k-vector is an element of $\Lambda_k\mathbb{E}^N$ and a k-covector (also called a k-form) is an element of $\Lambda^k\mathbb{E}^N$. A k-vector z (resp., k-covector z^*) is *simple* or *decomposable* if it can be written as a wedge product of k vectors (resp., k covectors):

$$z = v_1 \wedge \ldots \wedge v_k, v_i \in \mathbb{E}^N, 1 \leq i \leq k$$

(resp.,
$$z^* = v_1^* \wedge \ldots \wedge v_k^*, v_i^* \in \mathbb{E}^{N^*}, 1 \le i \le k).$$

If
$$e_1 = (1, 0, \ldots, 0), \ldots, e_N = (0, \ldots, 0, 1)$$

is the standard basis of $\mathbb{E}^N$, then a basis of $\Lambda_k \mathbb{E}^N$ is given by the decomposable k-vectors
$$e_{i_1} \wedge \ldots \wedge e_{i_k}, 1 \le i_i < \ldots < i_k \le N.$$

If $\theta^1, \ldots, \theta^N$ is the dual frame of $e_1, \ldots, e_N$, then a basis of $\Lambda^k \mathbb{E}^N$ is given by the decomposable k-covectors
$$\theta^{i_1} \wedge \ldots \wedge \theta^{i_k}, 1 \le i_1 < \ldots < i_k \le N.$$

In particular, $\Lambda_N \mathbb{E}^N$ and $\Lambda^N \mathbb{E}^N$ have dimension 1. The N-covector $\theta^1 \wedge \ldots \wedge \theta^N$ is a frame of $\Lambda^N \mathbb{E}^N$. It is called the *standard orientation* of $\mathbb{E}^N$.

For every $k, 1 \le k \le N$, the space $\Lambda^k \mathbb{E}^N$ is endowed with its classical operator norm, defined as follows: if $\omega \in \Lambda^k \mathbb{E}^N$,

$$|\omega| = \sup\{< \omega, z >, z \text{ simple } k\text{-vector of } \mathbb{E}^N\}.$$

An important remark. Using the standard scalar product $<,>$ on $\mathbb{E}^N$, we often identify a vector v with its corresponding 1-form $< v, . >$. Consequently, e_i is identified with θ^i and the N-vector $e_1 \wedge \ldots \wedge e_N$ is identified with the orientation $\theta^1 \wedge \ldots \wedge \theta^N$.

2. **The differential point of view.**

- *The space $\mathcal{D}^k(\mathbb{E}^N)$.* Let x be a point of $\mathbb{E}^N$. One can identify $T_x \mathbb{E}^N$ with $\mathbb{E}^N$ and $\Lambda^k T_x \mathbb{E}^N$ with $\Lambda^k \mathbb{E}^N$. In such a way, one constructs a vector bundle over $\mathbb{E}^N$ whose fiber over a point x is $\Lambda^k T_x E^N \simeq \Lambda^k \mathbb{E}^N$. A k-differential form on $\mathbb{E}^N$ is a C^{∞}[1] section of this bundle.

  Remark In particular, any N-differential form ω of $\mathbb{E}^N$ can be written

  $$\omega = f \omega_0,$$

  with $\omega_0 = \theta^1 \wedge \ldots \wedge \theta^N$, f being a C^{∞} real function.
  The space of k-differential forms on $\mathbb{E}^N$ is denoted by $\mathcal{D}^k(\mathbb{E}^N)$. It has a natural module structure on $C^{\infty}(\mathbb{E}^N)$.

  We endow $\mathcal{D}^k(\mathbb{E}^N)$ with the topology spanned by the family v_K^l of *seminorms* defined for every integer l and every compact subset K of $\mathbb{E}^N$ as follows: if

  $$\omega = \sum_{i_1 \ldots i_k} f_{(i_1 \ldots i_k)} \theta^{i_1} \wedge \ldots \wedge \theta^{i_k},$$

  then
  $$v_K^l(\omega) = \sup\{|d^j f_{(i_1 \ldots i_k)}(x)|; x \in K, 0 \le j \le l\}.$$

---

[1] In general, C^2 will be enough in this book.

- *The space $\mathcal{D}_c^k(\mathbb{E}^N)$.* Let us now make precise the notion of *support* of a differential form. If ω is a k-differential form, the *support* spt(ω) of ω is the closure of the set of points x such that $\omega_x \neq 0$. If K is a compact subset of $\mathbb{E}^N$, we denote by:

  – $\mathcal{D}_K^k(\mathbb{E}^N)$ the (closed) subspace of k-differential forms whose support is included in K, endowed with the induced topology.
  – $\mathcal{D}_c^k(\mathbb{E}^N)$ the union of $\mathcal{D}_K^k(\mathbb{E}^N)$ over all compact subsets K of $\mathbb{E}^N$, i.e., the subspace of k-differential forms with compact support. The space $\mathcal{D}_K^k(\mathbb{E}^N)$ is endowed with the largest topology such that the inclusion of each $\mathcal{D}_K^k(\mathbb{E}^N)$ into $\mathcal{D}_c^k(\mathbb{E}^N)$ is continuous.

  (The reader can consult [43, pp. 343–346] for details.)

- *The exterior derivative.* We denote by d the exterior derivative:

$$d : \mathcal{D}^k(\mathbb{E}^N) \to \mathcal{D}^{k+1}(\mathbb{E}^N).$$

The operator d satisfies

$$d^2 = 0.$$

- *Pullback of a differential form.* Let

$$f : \mathbb{E}^N \to \mathbb{E}^{N'}$$

be a differentiable map. Let ω be a k-differential form on $\mathbb{E}^{N'}$. One defines the k-differential form $f^*(\omega)$ on $\mathbb{E}^N$ (called the *pullback* of ω) as follows:

$$\forall X_1, ..., X_k \in T\mathbb{E}^N, f^*(\omega)(X_1, ..., X_k) = \omega(df(X_1), ..., df(X_k)). \qquad (8.1)$$

In particular, if $k = N = N'$, a simple computation shows that

$$f^*(\omega) = J(f)\omega, \qquad (8.2)$$

where $J(f)$ denotes the Jacobian of f.

The pullback commutes with the exterior derivative: for every map $f : \mathbb{E}^N \to \mathbb{E}^{N'}$ and every k-differential form ω on $\mathbb{E}^{N'}$, one has

$$f^*d\omega = d[f^*\omega]. \qquad (8.3)$$

8.1.2 Integration of N-Differential Forms on $\mathbb{E}^N$

Let $\mathbb{E}^N$ be the N-dimensional Euclidean space *endowed with its standard orientation* ω_0. We have seen that any N-form ω on $\mathbb{E}^N$ can be written as $\omega = f\omega_0$, where f is a smooth function on $\mathbb{E}^N$. If f is Lebesgue integrable (see Sect. 5.3), we put

$$\int \omega = \int f d\mathcal{L}^N.$$

This construction can be obviously extended to differential N-forms which are "less smooth," i.e., to differential N-forms of type $\omega = f\omega_0$ such that f is only integrable. Note, however, that "f positive" does not imply that $\int_A \omega$ is *positive*. Consequently, strictly speaking, the map

$$A \to \int_A f\omega_0$$

cannot induce a (positive) measure on $\mathbb{E}^N$.

8.2 Densities

8.2.1 Notion of Density on $\mathbb{E}^N$

As we have seen in Sect. 8.1.2, integration of differential forms on $\mathbb{E}^N$ does not allow us to define a measure on it: the result of the integration over any Borel subset may be negative. To avoid this difficulty, one introduces the notion of *density*:

- *Algebraic point of view*. A *density* on $\mathbb{E}^N$ is a map

$$\delta : \mathbb{E}^N \to \mathbb{R}^+,$$

  such that there exists a nonzero N-form $\alpha \in \Lambda^N(\mathbb{E}^N)$ satisfying

$$\delta(X_1, ..., X_N) = |\alpha(X_1, ..., X_N)|.$$

  There is a canonical density δ_0 on $\mathbb{E}^N$ associated to ω_0 (or equivalently, associated to the determinant with respect to an oriented orthonormal frame):

$$\delta_0(X_1, ..., X_N) = |\det(X_1, ..., X_N)| = \sqrt{\det < X_i, X_j >}.$$

  Clearly, two densities δ and δ' on $\mathbb{E}^N$ are proportional:

$$\delta' = a\delta, a \in \mathbb{R}^+.$$

- *Differential point of view*. We can extend this construction: a *(differential) density* on $\mathbb{E}^N$ is a map

$$\delta_x : T_x\mathbb{E}^N \to \mathbb{R},$$

  defined at each point x of $\mathbb{E}^N$, such that there exists a differential N-form, different to zero everywhere, such that

$$\forall x \in \mathbb{E}^N, \forall X_1, ..., X_N \in T_x\mathbb{E}^N, \delta_x(X_1, ..., X_N) = |\alpha_x(X_1, ..., X_N)|.$$

  Two densities δ and δ' on $\mathbb{E}^N$ are proportional:

$$\delta' = f\delta,$$

f being a positive function. In particular, any (differential) density δ can be written $\delta = f\delta_0$, where f is a positive function.

8.2.2 Integration of Densities on $\mathbb{E}^N$ and the Associated Measure

We can define the integral of a density with respect to the Lebesgue measure as follows. Let $\delta_f = f\delta_0$ be a density, f being a smooth positive function. We put

$$\int \delta_f = \int f d\mathcal{L}^N.$$

Note that this construction can be extended to the case where f is any integrable positive function. On the other hand, since f is positive, $\int \delta_f$ is nonnegative. Then, it is easy to check that the map

$$A \rightarrow \int_A \delta_f$$

is a new *Borel measure* on $(\mathbb{E}^N, \mathcal{B})$ (see Chap. 5).

Chapter 9
Measures on Manifolds

Let us introduce the concept of *measures* on a C^k-manifold M^n of dimension n with or without boundary ∂M^n ($n \geq 1, k \geq 2$). The goal is to use suitable differential forms to construct measures, with which one can define the notion of *volume*, fundamental in our context. Chapter 3 of [11] gives a complete introduction to the subject.

9.1 Integration of Differential Forms

We denote by $\mathcal{D}^k(M^n)$ the space of k-differential forms on M^n. If $\omega \in \mathcal{D}^k(M^n)$ and $x \in M^n$, ω_x is a k-form on the tangent space $T_x M^n$, i.e., $\omega_x \in \Lambda^k(T_x M^n)$.

A *volume form* on M^n is an n-differential form ω such that for every $x \in M^n$, $\omega_x \neq 0$. If ω is a volume form on M^n, the couple (M^n, ω) is an *oriented manifold*.

If (U, ϕ) is a chart domain of an oriented manifold (M^n, ω), one can define the integral $\int_U \omega$ of the restriction of ω to U as follows: $\phi^{-1^*}(\omega)$ is an n-form on $\phi(U)$, which is an open set of $\mathbb{E}^n$ endowed with its canonical orientation ω_0. We can write $\phi^{-1^*}(\omega) = f \omega_0$ and put

$$\int_U \omega = \int_{\phi(U)} f \, d\mathcal{L}^n, \tag{9.1}$$

where $d\mathcal{L}^n$ denotes the Lebesgue measure on $\mathbb{E}^n$. Formula (9.1) and Theorem 18 show that the result is independent of the chart. Using a partition of unity, one can then define $\int_A \omega$ in a natural way, where A is a subset of M^n such that $\phi(A)$ is measurable on $\mathbb{E}^n$. One more time, the map

$$A \to \int_A f \omega_0$$

cannot induce any (outer) measure on M^n, since the result of the integration may be negative.

Let us mention Stokes theorem, which is one of the most important result in mathematics.

Theorem 30. *Let M^n be a compact oriented C^k-manifold M^n of dimension n with boundary ∂M^n ($n \geq 1, k \geq 2$). Then, for any C^{k-1}-differential form ω of degree $n - 1$,*

$$\int_{M^n} d\omega = \int_{\partial M^n} \omega.$$

In particular, if $\partial M^n = \emptyset$,

$$\int_{M^n} d\omega = 0.$$

9.2 Density and Measure on a Manifold

We need to extend the notion of *density* we have introduced in Chap. 5 on $\mathbb{E}^N$. By definition, we say that δ is a density on M^n if at each point x of M^n, δ_x is a density of $T_x M^n$. Consequently, there exists an n-form ω on M^n such that $|\omega_x| = \delta_x$ at each point $x \in M^n$. In other words, $|\omega| = \delta$.

Theorem 31. *To any density on a C^k-manifold M^n, there is canonically associated a Borel measure on M^n. A measure on M^n obtained in such a way is called a Lebesgue measure.*

Remark Theorem 31 implies that, given a density δ on a manifold M^n, one can consider measurable subsets and integrable functions associated to the Borel measure defined with respect to δ. The main utility of densities is that they allow us to define *measures* on manifolds. Note that M^n is locally homeomorphic to $\mathbb{E}^n$ and then the Borel subsets of M^n are spanned by the images by this homeomorphism of the Borel subsets of $\mathbb{E}^n$.

Sketch of proof of Theorem 31 Let δ be a density on M^n:

- We begin by working on a domain U of a chart (U, ϕ). Let f be the function defined on U by

$$\phi^{-1*}(\delta) = f \delta_0,$$

where δ_0 is the canonical density on $\mathbb{E}^n$ (see Sect. 8.2). We define the measure μ_δ as follows: for any Borel subset A of U, we put

$$\mu_\delta(A) = \int_{\phi(A)} f d\mathcal{L}^n.$$

The main point is to check that this (local) definition of μ_δ is independent of the chart (we leave this proof to the reader).

- Then, we extend this construction to the whole manifold M^n by using a partition of unity. $\quad\square$

9.3 The Fubini Theorem on a Fiber Bundle

Among other ideas, the proof of the tubes formulas (given in Chap. 17) uses classically an "integration by slice," i.e., the *Fubini* theorem, to compute explicitly the volume of a tubular neighborhood of a submanifold. We give here (without proof) a general theorem on integration over a fiber bundle generalizing the integration by slice (see [51, 76] for instance).

Theorem 32. *Let* (E, π, B, F) *be an oriented fiber bundle over an oriented n-manifold B, such that* $\dim F = r$. *If E is endowed with the local product orientation induced by the orientation of B and F, then*

$$\int_E = \int_B \circ \int_F . \tag{9.2}$$

Equation (9.2) must be made more explicit. The assumptions of the theorem imply that the volume form ω_E of E is over each point p of B, "the product of the volume form ω_B by the volume form of the fiber ω_F." Roughly speaking, over each point ξ_m of E,

$$\omega_E = \omega_B \wedge \omega_F ,$$

where F_{ξ_m} denotes the fiber over m. Then, $\int_F \omega$ is a smooth function on B, which can be integrated, giving sense to $\int_B (\int_F \omega)$. *Fubini*'s theorem asserts that

$$\int_E \omega = \int_B (\int_F \omega).$$

Chapter 10
Background on Riemannian Geometry

We do not use much intrinsic Riemannian geometry in this book, since our manifolds are essentially submanifolds of Euclidean spaces. However, the Steiner formula for convex subsets with smooth boundaries and the tubes formula of Weyl (see Chaps. 16 and 17) are Riemannian results. Moreover, the invariant forms described in Chap. 19 have a typical Riemannian flavor. That is why we give a brief survey of Riemannian geometry. The reader interested in the subject can consult [10], [29], [76], [57, Tome 1, Chaps. 4 and 5], or [67].

10.1 Riemannian Metric and Levi-Civita Connexion

Let M^n be an n-dimensional manifold. We denote by TM^n its tangent bundle and by $\mathcal{X}(M^n)$ the module of C^∞ vector fields on M^n. A *Riemannian metric* on M^n is a positive definite symmetric tensor field g of type $(0,2)$ defined on M^n. In other words, at each point $m \in M^n$, g_m is a scalar product on $T_m M^n$ which varies differentiably with m. This tensor will often be denoted by $<,>$. Every manifold admits Riemannian metrics. To each Riemannian metric, $<,>$ is associated a unique linear connexion ∇, called the *Levi-Civita connexion*. Recall that a *linear connexion* ∇ on M^n is an operator

$$\nabla : \mathcal{X}(M^n) \times \mathcal{X}(M^n) \to \mathcal{X}(M^n),$$

$$(X,Y) \to \nabla_X Y,$$

satisfying, for all $X,Y,Z \in \mathcal{X}(M^n)$ and for all $f,g \in C^\infty(M^n)$:

- $\nabla_{fX+gY}Z = f\nabla_X Z + g\nabla_Y Z.$
- $\nabla_X(Y+Z) = \nabla_X Y + \nabla_X Z.$
- $\nabla_X fZ = f\nabla_X Z + X(f)Z.$

The torsion and the curvature are two tensors associated to any linear connexion:

- The torsion T is defined for all $X,Y \in \mathcal{X}(M^n)$ by

$$T(X,Y) = \nabla_X Y - \nabla_Y X - [X,Y].$$

- The curvature R is defined for all $X, Y, Z \in \mathcal{X}(M^n)$ by

$$R(X,Y)Z = \nabla_X \nabla_Y Z - \nabla_Y \nabla_X Z - \nabla_{[X,Y]} Z.$$

The Levi-Civita connexion is completely determined by two supplementary conditions:

1. It has no torsion: for all $X, Y \in \mathcal{X}(M^n)$,

$$\nabla_X Y - \nabla_Y X = [X,Y],$$

  where $[.,.]$ is the Lie bracket on M^n.
2. $<,>$ is parallel with respect to ∇: for all $X, Y, Z \in \mathcal{X}(M^n)$,

$$Z < X, Y > = < \nabla_Z X, Y > + < X, \nabla_Z Y > .$$

10.2 Properties of the Curvature Tensor

The curvature tensor R of ∇ satisfies the following algebraic properties. For all $X, Y, Z, W \in \chi(M^n)$:

- $R(X,Y)Z + R(Y,X)Z = 0$.
- $R(X,Y)Z + R(Y,Z)X + R(Z,X)Y = 0$, called the *first Bianchi identity*.
- $< R(X,Y)Z,W > + < R(Y,X)W,Z > = 0$.
- $< R(X,Y)Z,W > = < R(Z,W)X,Y > = 0$.

At each point $m \in M^n$ and each two-dimensional plane π of $T_m M^n$, one defines the *sectional curvature* $K(\pi)$ of π by

$$K(\pi) = < R(X,Y)Y,X >,$$

where (X,Y) is an orthonormal frame of π.

If X and Y are two vectors in $T_m M^n$, one defines the *Ricci tensor* Ricc of M^n for all $X, Y \in \chi(M^n)$ by

$$\text{Ricc}(X,Y) = \sum_{n=1}^{n} < R(e_i,X)Y, e_i >,$$

where $(e_1, ..., e_i, ..., e_n)$ is an orthonormal frame of $T_m M^n$.

The *scalar curvature* at m is the real number

$$\tau = \frac{1}{n(n-1)} \sum_{i=1}^{n} \text{Ricc}(e_i, e_i).$$

10.3 Connexion Forms and Curvature Forms

Let us now define the *connexion forms* on M^n. Let $(e_1, ..., e_i, ..., e_n)$ be a local ortho-normal frame of vector fields. Denote by $(\omega^1, ..., \omega^i, ..., \omega^n)$ the dual frame. The n^2 *connection 1-forms* ω_i^j are defined for all $X \in TM^n$ by

$$\nabla_X e_j = \sum_i \omega_j^i(X) e_i, \forall j \in \{1, ..., n\}.$$

Moreover, let R_{jkl}^i be the component of the curvature tensor:

$$R(e_j, e_k) e_l = \sum_i R_{ljk}^i e_i.$$

The curvature 2-forms of the connexion are defined by

$$\Omega_j^i = \frac{1}{2} \sum_{k,l} R_{jkl}^i \omega^k \wedge \omega^l.$$

The forms $\omega^i, \omega_i^j, \Omega_j^i$ are related by the *Cartan equations*:

$$d\omega^i = -\sum_j \omega_j^i \wedge \omega^j,$$

$$d\omega_j^i = -\sum_k \omega_k^i \wedge \omega_j^k + \Omega_j^i.$$

As usual, using Einstein summation, we avoid in general the sign $\sum$ when no confusion is possible. For instance,

$$\omega_j^i(X) e_i$$

must be read

$$\sum_i \omega_j^i(X) e_i.$$

10.4 The Volume Form

On any orientable Riemannian manifold M^n, the *volume form* dv is the n-form defined locally by $dv = \omega^1 \wedge ... \wedge \omega^i \wedge ... \wedge \omega^n$, where as before $(\omega^1, ..., \omega^i, ..., \omega^n)$ is the dual frame of an orthonormal local frame $(e_1, ..., e_i, ..., e_n)$.

Suppose that M^n is compact. The *volume* $\mathrm{Vol}_n(M^n)$ of M^n is defined by

$$\mathrm{Vol}_n(M^n) = \int_{M^n} dv.$$

If no confusion is possible, we simply write $\mathrm{Vol}(M^n)$ instead of $\mathrm{Vol}_n(M^n)$.

10.5 The Gauss–Bonnet Theorem

The famous *Gauss–Bonnet formula* gives a relation between the curvature tensor of
any even-dimensional Riemannian manifold M^n, $n = 2m$, and its topology. It can be
stated as follows. Let

$$\Omega = \frac{(-1)^m}{2^{2m}\pi^m m!} \sum \varepsilon_{i_1 \dots i_{2m}} \Omega_{i_2}^{i_1} \wedge \dots \wedge \Omega_{i_{2m}}^{i_{2m-1}}.$$

Then

$$\int_{M^n} \Omega = \chi(M^n), \tag{10.1}$$

where $\chi(M^n)$ denotes the Euler characteristic of M^n.

10.6 Spheres and Balls

Spheres and balls are well-known compact Riemannian manifolds. Their volumes
can be expressed as follows. Let Γ be the gamma function. It is well known that Γ
satisfies the following property: for every $n \in \mathbb{N}$,

$$\Gamma(n+1) = n\Gamma(n); \; \forall n \in \mathbb{N}, \Gamma(n) = (n-1)!$$

and

$$\Gamma(\frac{1}{2}) = \sqrt{\pi}.$$

- We denote by $\mathbb{S}^n$ the unit hypersphere (of dimension n) in $\mathbb{E}^{n+1}$. Its n-volume s_n
  satisfies

$$s_n = \frac{2\pi^{\frac{(n+1)}{2}}}{\Gamma((\frac{(n+1)}{2}))}. \tag{10.2}$$

- We denote by $\mathbb{B}^n$ the unit ball of $\mathbb{E}^n$ (whose boundary is $\mathbb{S}^{n-1}$). Its n-volume b_n
  satisfies

$$b_n = \frac{s_n}{n} = \frac{2\pi^{\frac{n}{2}}}{n\Gamma((\frac{n}{2}))}. \tag{10.3}$$

In particular,

$$s_0 = 2; s_1 = 2\pi; s_2 = 4\pi; s_3 = 2\pi^2; b_1 = 2; b_2 = \pi; b_3 = \frac{4}{3}\pi. \tag{10.4}$$

10.7 The Grassmann Manifolds

10.7.1 The Grassmann Manifold $G^o(N,k)$

Let $G^o(N,k)$ be the set of (*oriented*) *k-linear* subspaces of $\mathbb{E}^N$. This set is called the *Grassmann manifold or the Grassmannian* of (oriented) *k*-vector subspaces (also called *k*-planes) of $\mathbb{E}^N$. It is a compact manifold, which has the structure of symmetric space of dimension $k(N-k)$ isomorphic to the quotient

$$\frac{SO(N)}{SO(k) \times SO(N-k)},$$

where $SO(N)$ (resp., $SO(k)$, resp., $SO(N-k)$) denotes the special orthogonal group of $\mathbb{E}^N$ (resp., $\mathbb{E}^k$, resp., $\mathbb{E}^{N-k}$).

Note that there is an obvious identification

$$G^o(N,k) \sim G^o(N,n-k)$$

(by identifying a *k*-plane with its orthogonal) and that $\mathbb{S}^{N-1}$ can be identified with $G^o(N,1)$ (by identifying an oriented line with its unit vector).

As usual, the tangent space $T_{\mathrm{id}}G^o(N,k)$ of $G^o(N,k)$ at the identity can be identified with

$$\frac{\mathcal{SO}(N)}{\mathcal{SO}(k) \times \mathcal{SO}(N-k)},$$

where $\mathcal{SO}(N)$ (resp., $\mathcal{SO}(k)$, resp., $\mathcal{SO}(N-k)$) denotes the space of skew-symmetric squared $N \times N$ (resp., $k \times k$, resp., $(N-k) \times (N-k)$) matrices. Consequently, a tangent vector at the identity can be identified with a matrix

$$\begin{pmatrix} 0 & A \\ -A^t & 0 \end{pmatrix}, \tag{10.5}$$

where A describes the space of $k \times (N-k)$ real matrices.

The manifold $G^o(N,k)$ has a Riemannian structure, which can be described by giving explicitly the associated scalar product $<,>$ on the tangent space $T_{\mathrm{id}}G^o(N,k)$ at *Id*: if $U = \begin{pmatrix} 0 & A \\ -A^t & 0 \end{pmatrix}$ and $U' = \begin{pmatrix} 0 & B \\ -B^t & 0 \end{pmatrix}$ belong to $G^o(N,k)$, then

$$<U,U'> = 2\mathrm{trace}\, AB^t,$$

i.e. (up to the coefficient 2), the standard scalar product of A and B.

The manifold $G^o(N,k)$ admits a *Haar measure* (invariant by rotations), associated to the volume form induced by the Riemannian metric. The volume of $G^o(N,n-k)$ satisfies

$$\mathrm{Vol}(G^o(N,k)) = \frac{2s_{N-1}s_{N-2}\ldots s_k}{s_{N-k-1}s_{N-k-2}\ldots s_0}. \tag{10.6}$$

In particular, the volume of the set of *(oriented)* lines through the origin in $\mathbb{E}^2$ (i.e., the volume of $\mathbb{S}^1$) is 2π; the volume of the set of *(oriented)* lines through the origin in $\mathbb{E}^3$ (i.e., the volume of the sphere $\mathbb{S}^2$) is 4π.

10.7.2 The Grassmann Manifold G(N, k)

We now deal with *unoriented* k-subspace of E^N. The set $G(N,k)$ of (unoriented) k-vector subspaces of $\mathbb{E}^N$ is called the Grassmann manifold *or Grassmannian* of (unoriented) k-planes of $\mathbb{E}^N$. Like $G^o(N,k)$, $G(N,k)$ has the structure of a differentiable manifold of dimension $k(N-k)$. The main observation is that the map

$$\psi : G^o(N,k) \to G(N,k), \tag{10.7}$$

which sends an *oriented* k-plane to the corresponding *unoriented* k-plane, is a twofold covering.

$G(N,k)$ is also a compact manifold, which has the structure of symmetric space of dimension $k(N-k)$ isomorphic to

$$\frac{O(N)}{O(k) \times O(N-k)}.$$

There is still an obvious identification:

$$G(N,k) \sim G(N,n-k)$$

(by identifying a k-plane with its orthogonal). Note also that $G(N,1)$ is the projective space $P^{N-1}(R)$.

The tangent space $T_{\mathrm{id}}G(N,k)$ at identity is still identified with

$$\frac{\mathcal{SO}(N)}{\mathcal{SO}(k) \times \mathcal{SO}(N-k)}.$$

The manifold $G(N,k)$ is endowed with a Riemannian structure defined by claiming that the map ψ is a local isometry. The corresponding volume is half the volume of $G^o(N,k)$:

$$\mathrm{Vol}(G(N,k)) = \frac{s_{N-1}s_{N-2}\ldots s_k}{s_{N-k-1}s_{N-k-2}\ldots s_0}. \tag{10.8}$$

In particular, the volume of the set of *(unoriented)* lines through the origin in $\mathbb{E}^2$ (i.e., the volume of the projective space $P^1(\mathbb{R})$) is π; the volume of the set of *(unoriented)* lines through the origin in $\mathbb{E}^3$ (i.e., the volume of the projective space $P^2(\mathbb{R})$) is 2π; and the volume of the set of *(unoriented)* planes through the origin in $\mathbb{E}^3$ is 2π.

10.7.3 The Grassmann Manifolds AG(N, k) and AG^o(N, k)

Let us now consider the set $AG(N,k)$ of all *affine k*-subspaces of $\mathbb{E}^N$. It is a (smooth) manifold with fibers over $G(N, N-k)$

$$\phi : AG(N,k) \to G(N, N-k),$$

where the map ϕ sends any affine k-subspace P onto its linear orthogonal complement L. The fibers are isomorphic to $\mathbb{E}^N$. The map

$$P \to L$$

gives the isomorphism of the fiber with $\mathbb{E}^N$. The manifold $AG(N,k)$ admits a natural canonical Haar measure (invariant under rigid motion), deduced from the Lebesgue measure on the Euclidean space and the Haar measure on $G(N, N-k)$ (see Theorem 32 and [56] for details).

One can also define the manifold $AG^o(N,k)$ of *oriented k*-subspaces of $\mathbb{E}^N$. The details of the corresponding fibration over $G^o(N, N-k)$ and the Haar measure (invariant under rigid motions) are left to the reader.

Chapter 11
Riemannian Submanifolds

Smooth Riemannian submanifolds in Euclidean spaces are the smooth objects, on which we shall test the *curvature measures* defined in the next chapters. They are the direct generalization in any dimension and codimension of curves and surfaces in $\mathbb{E}^3$. Their extrinsic curvatures generalize the Gauss and mean curvatures of surfaces. We review (without proof) some fundamental notions on the subject. Classical books on Riemannian submanifolds are [26, 27].

11.1 Some Generalities on (Smooth) Submanifolds

Let M^n be a (smooth) n-dimensional manifold (in general, if there is no other condition, we assume that M^n is at least C^2). A C^1-map

$$x : M^n \to \mathbb{E}^N$$

is an *immersion* if dx has maximal rank n at each point. This immersion is an *embedding* if

$$x : M^n \to x(M^n)$$

is an homeomorphism (for the topology on $x(M^n)$ induced by the one on $\mathbb{E}^N$). In such a situation, we say that M^n is an immersed (or embedded) *submanifold* of $\mathbb{E}^N$, or simply a submanifold of $\mathbb{E}^N$ (Fig. 11.1).

Let us endow $\mathbb{E}^N$ with its standard scalar product $<,>$ and denote by $g = x^*(<,>)$ the induced metric on M^n. Then, x becomes an *isometric immersion* of (M^n, g) into $(\mathbb{E}^N, <,>)$. In the following, we identify systematically a tangent vector X to M^n and its image $x_* X$ by the differential of x. Using this identification, the orthogonal complement of TM^n in $T\mathbb{E}^N$ is the normal bundle of M^n in $\mathbb{E}^N$. We denote it by $T^\perp M^n$, and by $\mathcal{X}^\perp(M^n)$ the module of normal vector fields. Consequently, one can consider the following three bundles over M^n (Fig. 11.2). The tangent bundle to M^n:

$$TM^n \to M^n,$$

Fig. 11.1 The curve C is the image of an immersion with a self-intersection point; this immersion is not an embedding. The curve C' is the image of an embedding

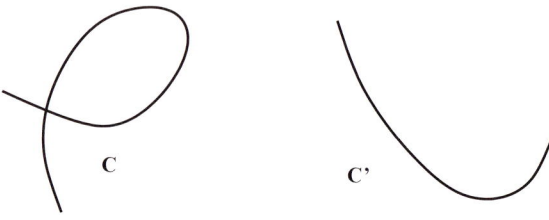

C

C'

Fig. 11.2 The tangent space T_pM and the normal space $T_p^\perp M$ of M at a point p

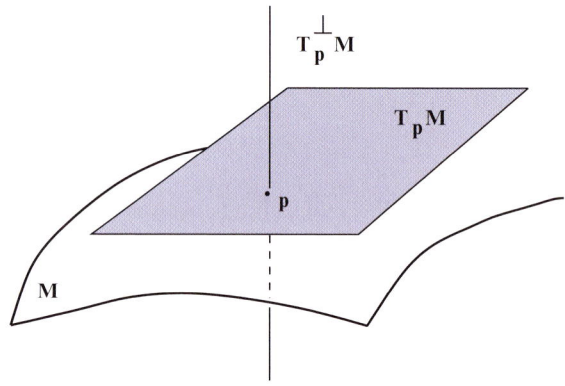

the restriction to M^n of the tangent bundle of $\mathbb{E}^N$:

$$T_{M^n}\mathbb{E}^N \to M^n,$$

and the normal bundle to M^n:

$$T^\perp M^n \to M^n.$$

These three bundles are related by the equation:

$$T_{M^n}\mathbb{E}^N = TM^n \oplus T_{M^n}^\perp M^n. \tag{11.1}$$

Among others, there are two important topological properties of the normal bundle of submanifolds, which will be used extensively in this book:

1. The first one is topological. It is the *tubular neighborhood theorem* (see [58, 62] for instance).

   Theorem 33. *Let M^n be a compact embedded submanifold of $\mathbb{E}^N$. Then, there exists a neighborhood U_{M^n} of M^n, such that every point m in U_{M^n} has a unique projection $\mathrm{pr}_{M^n}(m)$ onto M^n (realizing the distance $\mathrm{d}_{M^n}(m)$ from m to M^n).*

   Since M^n is compact, Theorem 33 implies that there exists a tubular neighborhood M_ε^n of M^n (of radius $\varepsilon > 0$), such that every m in M_ε^n has a unique projection $\mathrm{pr}_{M^n}(m)$ onto M^n. In other words, using the notion of *reach* of a subset, introduced in Chap. 4, one has the following corollary.

Fig. 11.3 In the tubular neighborhood U_C of the curve C in $\mathbb{E}^3$, any point m has a unique orthogonal projection onto C

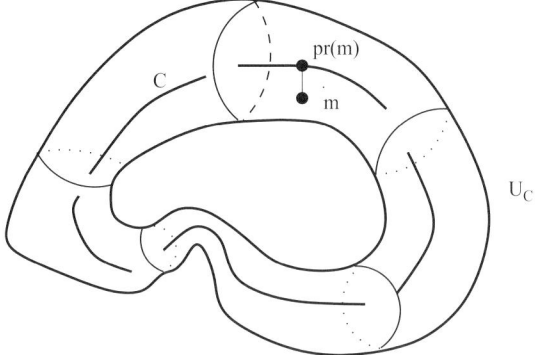

Corollary 1. *The reach of a compact submanifold of $\mathbb{E}^N$ is strictly positive (Fig. 11.3).*

Note that the smoothness of the immersion is essential. Theorem 33 and Corollary 1 fail for a polygonal line in $\mathbb{E}^2$, as we have seen in Chap. 4: any point on the bisector of two consecutive edges has at least two orthogonal projections.

2. The second one comes from symplectic geometry (see [81] for a nice introduction). Remember that the cotangent bundle of any N-dimensional manifold V^N is a manifold which admits a canonical *symplectic* 2-form Ω (i.e., a closed differential 2-form Ω, which is nondegenerate and satisfies

$$\Omega^N \neq 0$$

at every point). In particular, if $(x_1, ..., x_k, ..., x_N)$ are the standard coordinates of $\mathbb{E}^N$ and $(x_1, ..., x_k, ..., x_N, y_1, ..., y_k, ..., y_N)$ are the standard coordinates of $T^*\mathbb{E}^N = \mathbb{E}^N \times \mathbb{E}^{*N}$, the canonical symplectic form Ω of $T^*\mathbb{E}^N$ has the following expression:

$$\Omega = \sum_{k=1}^{N} dx_k \wedge dy_k. \tag{11.2}$$

Using the identification given by the metric between vectors and 1-forms, we deduce a symplectic structure (still denoted by Ω) on the tangent bundle $T\mathbb{E}^N = \mathbb{E}^N \times \mathbb{E}^N$ of $\mathbb{E}^N$.

It is well known that Ω is exact

$$\Omega = d\alpha, \tag{11.3}$$

where α is the *Liouville* 1-form

$$\alpha = -\sum_{k=1}^{N} y_k dx_k. \tag{11.4}$$

Let us now restrict our attention to the unit sphere bundle

$$ST\mathbb{E}^N = \mathbb{E}^N \times \mathbb{S}^{N-1}$$

of $T\mathbb{E}^N$. The restriction of the Liouville form α to $\mathbb{E}^N \times \mathbb{S}^{N-1}$ endows $\mathbb{E}^N \times \mathbb{S}^{N-1}$ with a *contact structure*: still denoting this restriction by α, one has

$$\alpha \wedge (d\alpha)^{N-1} \neq 0 \qquad (11.5)$$

at every point (x, y) of $\mathbb{E}^N \times \mathbb{S}^{N-1}$.

In this context, let us describe the behavior of the normal (resp., unit normal) bundle of a submanifold M^n of $\mathbb{E}^N$. A simple direct computation implies the following theorem.

Theorem 34. *Let M^n be a submanifold of $\mathbb{E}^N$. Then:*

- *Considered as a submanifold of $\mathbb{E}^N \times \mathbb{E}^N$ endowed with the symplectic structure Ω, the normal bundle $T^\perp M^n$ of M^n is Lagrangian, i.e.,*

$$\forall X, Y \in TT^\perp M^n, \Omega(X, Y) = 0.$$

- *Considered as a submanifold of $\mathbb{E}^N \times \mathbb{S}^{N-1}$, the unit normal bundle $ST^\perp M^n$ of M^n is Legendrian, i.e.,*

$$\forall X, Y \in TST^\perp M^n, \alpha(X) = 0 \text{ and } d\alpha(X, Y) = 0.$$

Theorem 34 is crucial for our purpose: in Chap. 20, following [47], the normal cycle of a geometric subset of $\mathbb{E}^N$ will satisfy analogous properties *by definition*.

11.2 The Volume of a Submanifold

The volume of a submanifold generalizes the length of a curve and the area of a surface. It only involves the first derivative of the immersion. Let M^n be a (compact orientable) C^1-submanifold of $\mathbb{E}^N$. There are (at least) three classical ways to define the n-volume of M^n:

1. We can endow M^n with the Riemannian structure induced by the scalar product on $\mathbb{E}^N$ and then consider the Lebesgue measure of M^n associated to the corresponding density.
2. We can compute the nth-Hausdorff measure of M^n, considered as a subset of $\mathbb{E}^N$.
3. If the submanifold is the image of an embedding

$$x : M^n \to \mathbb{E}^N,$$

its volume can be computed with help of the area formula (Theorem 19):

$$\mathrm{Vol}_n(M^n) = \int_{M^n} J_n(f)(x)d\mathcal{L}^n x. \qquad (11.6)$$

It can be proved that these three computations coincide. They define the n-volume $\mathrm{Vol}_n(M^n)$ of M^n. Consequently,

$$\mathrm{Vol}_n(M^n) = \mathcal{H}^n(M^n) = \int_{M^n} J_n(f)(x)d\mathcal{L}^n x. \qquad (11.7)$$

Remark In this section, we supposed that the submanifold is C^1. However, a sharp study of these processes shows that the same considerations can be applied to submanifolds defined with Lipschitz maps, since such maps are differentiable almost everywhere, see [43, 63] for details.

11.3 Hypersurfaces in $\mathbb{E}^N$

Let us now deal with the local *Riemannian* geometry of submanifolds. We begin with hypersurfaces of $\mathbb{E}^N$ (see [26] for details). Although we can mimic Sect. 20.2 to define the Weingarten endomorphism and the second fundamental form of a hypersurface, we shall be a little more formal, using the *covariant derivative* in the Euclidean space (i.e., the Levi-Civita connexion in the Euclidean space $\mathbb{E}^N$, defined in a general frame in Sect. 10.1). Let $(e_1, ..., e_i, ..., e_N)$ be an orthonormal frame of $\mathbb{E}^N$. The *covariant derivative* on $\mathbb{E}^N$ is simply the operator $\tilde{\nabla}$ defined as follows. If

$$X = \sum_{i=1}^{N} X^i e_i, \; Y = \sum_{j=1}^{N} Y^j e_j$$

are two vector fields on $\mathbb{E}^N$, then

$$\tilde{\nabla}_X Y = \sum_{i,j=1}^{N} X^i \frac{\partial Y^j}{\partial x^i} e_j.$$

11.3.1 The Second Fundamental Form of a Hypersurface

In our context, an (oriented) *smooth hypersurface M* of the (oriented) Euclidean space $\mathbb{E}^N$ means an $(N-1)$-dimensional C^2-(oriented) manifold embedded in $\mathbb{E}^N$. Let ξ be the unit normal vector field defined on M, compatible with the orientation of M and $\mathbb{E}^N$. The Weingarten tensor A defined on M is the symmetric endomorphism defined for all $X \in TM$ by

$$A(X) = -\tilde{\nabla}_X \xi. \tag{11.8}$$

The *second fundamental form h* of M is the adjoint of A: it is defined for all $X, Y \in TM$ by

$$< A(X), Y > = h(X, Y). \tag{11.9}$$

The geometric meaning of h and A is well known: it describes the *shape* of the immersion.[1] Since A is symmetric, it admits at each point $m \in M$ an orthonormal frame of eigenvectors called *principal vectors*, whose corresponding eigenvalues are called the *principal curvatures* $(\lambda_1, ..., \lambda_i, ..., \lambda_{N-1})$ of M at m.

11.3.2 k^{th}-Mean Curvature of a Hypersurface

Let us now introduce the symmetric functions of the principal curvatures.

Definition 25. Let M be a hypersurface of $\mathbb{E}^N$. For every, $k = 0, ..., N-1$, the kth-elementary symmetric function of the principal curvatures of M

$$\Xi_k = \{\lambda_{i_1}, ..., \lambda_{i_k}\}$$

is called the kth-mean curvature of M.

Remark One has

$$\det (I + tA) = \sum_{k=0}^{k=N-1} \Xi_k t^k, \tag{11.10}$$

with $\Xi_k = \sum_{I_k} \Delta_{I_k}$, where Δ_{I_k} is the sum of all k-minors of the matrix A (I_k denotes the class of subsets of $\{1, ..., N-1\}$ with k elements). In particular:

- $\Xi_0 = 1$.
- Ξ_1 is the trace of h.
- Ξ_{N-1} is the determinant G of h, called the *Gauss curvature* of M.

In particular, we have the following result.

Definition 26. The scalar

$$H = \frac{1}{N-1} \Xi_1$$

is called the mean curvature of M.

---

[1] For instance, M is totally geodesic in $\mathbb{E}^N$ (i.e., $h = 0$), if the geodesics of M^n are sent onto geodesics (i.e., straight lines) of $\mathbb{E}^N$; in this case, $x(M)$ is nothing but a portion of an affine hyperplane of $\mathbb{E}^N$.

11.4 Submanifolds in $\mathbb{E}^N$ of Any Codimension

Section 11.3 can be generalized to any submanifold

$$x : M^n \to \mathbb{E}^N$$

of any codimension (as for Sect. 11.3, see [26] for details).

11.4.1 The Second Fundamental Form of a Submanifold

Let ∇ (resp., $\tilde{\nabla}$) denote the Levi-Civita connexion on (M^n, g) (resp., $(\mathbb{E}^N, <,>)$). For all vector fields X, Y tangent to M^n, $\tilde{\nabla}_X Y$ can be considered as a section of $T_{M^n}\mathbb{E}^N$. It appears that its component tangent to M^n is exactly $\nabla_X Y$, so that the two connections ∇ and $\tilde{\nabla}$ are related by the *Gauss formula*, for all $X, Y \in \chi(M^n)$,

$$\tilde{\nabla}_X Y = \nabla_X Y + h(X, Y), \tag{11.11}$$

where $h(X, Y) \in T^\perp M^n$ denotes the normal component of $\tilde{\nabla}_X Y$. The tensor

$$h : TM^n \times TM^n \to T^\perp M^n$$

is symmetric. It is called the *second fundamental form* of M^n with respect to the immersion x.

Let ξ be a normal vector field and X be a tangent vector field on M^n. We can decompose $\tilde{\nabla}_X \xi$ into its tangent component $-A_\xi X$ and its normal component $\nabla_X^\perp \xi$. We get the *Weingarten formula*:

$$\tilde{\nabla}_X \xi = -A_\xi X + \nabla_X^\perp \xi, \forall X \in TM^n. \tag{11.12}$$

The tensor A is called the *Weingarten tensor* and is related to h by the following equation, for all $X, Y \in \chi(M^n)$ and for all $\xi \in T^\perp M^n$,

$$< A_\xi X, Y > = < h(X, Y), \xi >, \forall X, Y \in TM^n. \tag{11.13}$$

Since h is symmetric, we can define its trace with respect to orthonormal frames in TM^n.

Definition 27. The mean curvature vector field of the submanifold M^n is the normal vector field $\overrightarrow{H}$ defined by

$$\overrightarrow{H} = \frac{1}{n} \, \text{trace} \, (h).$$

11.4.2 k^{th}-Mean Curvatures in Large Codimension

Generalizing Definition 28, one defines for each $k \in \{0,...,n\}$ the kth-mean curvature with respect to each normal direction as follows. Let ξ be a (unit) normal vector field. Let

$$h^\xi : TM^n \times TM^n \to C^\infty(M^n)$$

be the tensor defined for all $X, Y \in \mathcal{X}(M^n)$ and for all $\xi \in T^\perp M^n$ by

$$h^\xi(X,Y) = <h(X,Y), \xi> . \tag{11.14}$$

In other words, h^ξ is the second fundamental form in the direction ξ. Since h is symmetric, A_ξ is self-adjoint and h^ξ is symmetric. Then, they can be (locally) diagonalized (in an orthonormal frame). At each point m of M^n and for any normal direction ξ at m, the eigenvalues $(\lambda_1^\xi,...,\lambda_n^\xi)$ of A_ξ are called *the principal curvatures* of M^n at m in the direction ξ.

Definition 28. Let M^n be an n-dimensional Riemannian manifold isometrically immersed in $\mathbb{E}^N$. Let ξ be a (unit) normal vector field. The kth-elementary symmetric function of the principal curvatures of M^n in the direction ξ

$$\Xi_k(\xi) = \{\lambda_{i_1}^\xi,...,\lambda_{i_k}^\xi\} \tag{11.15}$$

is called the kth-mean curvature of M^n in the direction ξ.

Remarks

- One has

$$\det(I + tA_\xi) = \sum_{k=0}^{k=n} \Xi_k(\xi)t^k, \tag{11.16}$$

with $\Xi_k(\xi) = \sum_{I_k} \Delta_{I_k}(\xi)$, where $\Delta_{I_k}(\xi)$ is the sum of all k-minors of the matrix A_ξ (I_k denotes the class of subsets of $\{1,...,n\}$ with k elements).
- If we replace ξ by $-\xi$, Ξ_k is changed into $-\Xi_k$ when k is odd, and is unchanged when k is even.

11.4.3 The Normal Connexion

The normal component

$$\nabla^\perp : \mathcal{X}(M^n) \times \mathcal{X}^\perp(M^n) \to \mathcal{X}^\perp(M^n)$$

of the Levi-Civita connexion $\tilde{\nabla}$ defined by (11.12) is a linear connection on the normal bundle $\chi^\perp(M^n)$: it satisfies for all $X, Y \in \mathcal{X}(M^n)$, for all $\xi, \zeta \in \mathcal{X}^\perp(M^n)$, and for all $f, g \in C^\infty(M^n)$:

- $\nabla^\perp_{fX+gY}\xi = f\nabla^\perp_X\xi + g\nabla^\perp_Y\xi$.
- $\nabla^\perp_X(\xi + \zeta) = \nabla^\perp_X\xi + \nabla^\perp_X\zeta$.
- $\nabla^\perp_X f\xi = f\nabla^\perp_X\xi + X(f)\xi$.

The operator $\nabla^\perp$ is called the *normal connexion* associated to the submanifold M^n.

The *normal curvature tensor* $R^\perp$ of the connexion $\nabla^\perp$ is defined as follows, for all $X, Y \in \mathcal{X}(M^n)$ and for all $\xi \in \mathcal{X}^\perp(M^n)$,

$$R^\perp(X,Y)\xi = \nabla_X\nabla_Y\xi - \nabla_Y\nabla_X\xi - \nabla_{[X,Y]}\xi.$$

11.4.4 The Gauss–Codazzi–Ricci Equations

The *Gauss* equation relates the curvature tensor R of M^n to the second fundamental form of the submanifold M^n. This equation can be stated as follows, for all $X, Y, Z, W \in TM^n$,

$$\begin{aligned} < R(X,Y)Z, W > = &< h(X,W), h(Y,Z) > \\ &- < h(X,Z), h(Y,W) > . \end{aligned} \tag{11.17}$$

The *Codazzi* equation is obtained by expressing the fact that the normal component of the curvature tensor of $\mathbb{E}^N$ is null, for all $X, Y, Z, W \in TM^n$,

$$(\overline{\nabla}_X h)(Y,Z) - (\overline{\nabla}_Y h)(X,Z) = 0, \forall X, Y, Z \in TM^n, \tag{11.18}$$

where, for all $X, Y, Z \in \mathcal{X}(M^n)$,

$$\begin{aligned} (\overline{\nabla}_X h)(Y,Z) = &\nabla^\perp_X h(Y,Z) \\ &- h(\nabla_X Y, Z) - h(Y, \nabla_X Z). \end{aligned} \tag{11.19}$$

Finally, the *Ricci* equation relates the curvature tensor $\nabla^\perp M^n$ of the normal bundle $T^\perp M^n$ to the Weingarten tensor: once more, it is obtained by expressing the fact that the normal component of the curvature tensor of $\mathbb{E}^N$ is null, for all $X, Y \in TM^n$ and for all $\xi, \eta \in T^\perp M^n$,

$$\forall X, Y \in TM^n, < R^\perp(X,Y)\xi, \eta > = < [A_\xi, A_\eta](X), Y >, \tag{11.20}$$

where

$$[A_\xi, A_\eta] = A_\xi A_\eta - A_\eta A_\xi. \tag{11.21}$$

11.5 The Gauss Map of a Submanifold

We have defined in Chaps. 2 and 3 the Gauss map of a curve and a surface. This can be generalized to any submanifold in Euclidean spaces. Let us begin with hypersurfaces.

11.5.1 The Gauss Map of a Hypersurface

Let

$$x : M^{N-1} \to \mathbb{E}^N$$

be a codimension one immersion of an (oriented) manifold M^{N-1} into $\mathbb{E}^N$. The Gauss map associated to x is the map

$$\mathbf{G} : M^{N-1} \to \mathbb{S}^{N-1}$$

defined as follows: if ξ is the oriented normal vector field of M^{N-1}, we put for all $m \in M^{N-1}$

$$\mathbf{G}(m) = \xi_m.$$

If $dv_{\mathbb{S}^{N-1}}$ is the volume form of $\mathbb{S}^{N-1}$, one has

$$\mathbf{G}^*(dv_{\mathbb{S}^{N-1}}) = Gdv_{M^{N-1}}, \tag{11.22}$$

where $G = \Xi_{N-1}$ is the Gauss curvature function on M^{N-1}, i.e., the determinant of its second fundamental form (see (11.10)).

If M^{N-1} is closed and has an even dimension $N-1 = 2m$, *Gauss* equation (11.17) and *Gauss–Bonnet* theorem (10.1) imply

$$\int_{M^{2m}} Gdv_M = \frac{s_{2m}}{2} \chi(M^{2m}), \tag{11.23}$$

a crucial formula already mentioned in Chap. 10.

The interested reader can find generalizations of *Fenchel–Milnor* theorem 3 in the beautiful articles of Chern and Lashof [31, 32] and an extended study of this type of results in the book of Chen [27].

11.5.2 The Gauss Map of a Submanifold of Any Codimension

Let $x : M^n \to \mathbb{E}^N$ be an immersion of a manifold M^n into $\mathbb{E}^N$. The *Gauss map*

$$\mathbf{G} : M^n \to G^0(N, N-n)$$

sends each point $m \in M^n$ to its normal space $T_m^\perp M^n$, considered as a point of the Grassmann manifold $G^0(N, N-n)$.[2] A slight extension of this definition is to consider the map

$$\tilde{\mathbf{G}} : M^n \to \mathbb{E}^N \times G^0(N, N-n),$$

defined by

$$\tilde{\mathbf{G}}(m) = (m, T_m M^n).$$

This modification will be interesting when we shall deal with the *normal cycle* of a smooth submanifold. In the following, if no confusion is possible, we use indistinctly $\mathbf{G}$ or $\tilde{\mathbf{G}}$, calling systematically $\mathbf{G}$ the considered map.

Let us explicit the local expression of $\mathbf{G}$: if $(e_1, ..., e_n, e_{n+1}, ..., e_N)$ is a direct local frame at a point of M^n, such that $(e_1, ..., e_n)$ is tangent to M^n and $(e_{n+1}, ..., e_N)$ is normal to M^n, then one can write

$$\mathbf{G}(m) = e_{n+1} \wedge ... \wedge e_N.$$

Differentiating $\mathbf{G}$, one finds immediately

$$d\mathbf{G}_m(X) = \sum_i e_{n+1} \wedge ... \wedge e_{n+i-1} \wedge A_{e_{n+i}}(X) \wedge e_{n+i+1} \wedge ... \wedge e_N, \qquad (11.24)$$

where A denotes the Weingarten endomorphism of M^n. This indicates that the pullback of the volume form of the Grassmann manifold gives informations on the shape of M^n. This a priori anodyne remark will be useful when we shall extend the notion of *curvature* to a large class of objects.

---

[2] For technical reasons, it may be more interesting to modify the definition by sending each point m to its *tangent* space. Both constructions are equivalent.

Chapter 12
Currents

As we shall see in Chap. 20, the *normal cycles* associated to *geometric subsets* of $\mathbb{E}^N$ are integral currents, which are a particular type of rectifiable currents. We give here a short introduction to this subject. We end this chapter with important theorems used in the approximation and convergence results proved in the succeeding parts of the book. A nice introduction to this subject can be found in [63].

12.1 Basic Definitions and Properties on Currents

In measure theory, *currents* appear as a generalization of smooth submanifolds. Currents are dual to differential forms. We have defined the topological space $\mathcal{D}_c^k(\mathbb{E}^N)$ in Sect. 8.1.1(2).

Definition 29.

1. The space $\mathcal{D}_k(\mathbb{E}^N)$ of k-currents of $\mathbb{E}^N (0 \leq k \leq N)$ is the topological dual of the space $\mathcal{D}_c^k(\mathbb{E}^N)$ of k-differential forms with compact support on $\mathbb{E}^N$. The duality bracket will be denoted by $< .,. >$.
2. The support spt(T) of a k-current T of $\mathbb{E}^N$ is the smallest closed subset $C \subset \mathbb{E}^N$ such that, if $\omega \in \mathcal{D}_c^k(\mathbb{E}^N)$ satisfies spt$(\omega) \cap C = \emptyset$, then $< T, \omega >= 0$.

The space $\mathcal{D}_k(\mathbb{E}^N)$ is naturally endowed with the weak topology: if $(T_p)_{p \in \mathbb{N}}$ is a sequence of k-currents of $\mathbb{E}^N$ and T is a k-current of $\mathbb{E}^N$, then

$$\lim_{p \to \infty} T_p = T \iff \forall \omega \in \mathcal{D}_c^k(\mathbb{E}^N), \lim_{p \to \infty} < T_p, \omega >=< T, \omega > .$$

Let us now define the *boundary* of a current.

Definition 30.

- Every k-current T of $\mathbb{E}^N$ is associated a $(k-1)$-current ∂T of $\mathbb{E}^N$, called the boundary of T, defined as follows:

$$\forall \omega \in \mathcal{D}_c^{k-1}(\mathbb{E}^N), < \partial T, \omega >=< T, d\omega > .$$

- A cycle is a current with null boundary.

Classical Examples

1. Let

$$f : \mathbb{E}^N \to \mathbb{E}^{N'}$$

be a smooth map and let $T \in \mathcal{D}_k(\mathbb{E}^N)$. One defines the k-current

$$f_*(T) \in \mathcal{D}_k(\mathbb{E}^{N'})$$

by the following formula, for every k-differential form $\omega \in \mathcal{D}_c^k(\mathbb{E}^{N'})$,

$$< f_*(T), \omega >=< T, f^*(\omega) > .$$

2. Let M^n be a compact smooth oriented submanifold of $\mathbb{E}^N$. Then, M^n can be considered as an n-current. Indeed, let dv_{M^n} be the volume form of M^n. If ω is any differential n-form on $\mathbb{E}^N$, then there exists a smooth function f_ω on M^n such that the restriction of ω on M^n equals $f_\omega dv_{M^n}$. The current $[M^n]$ associated to M^n is defined as follows. For all $\omega \in \mathcal{D}_c^k(\mathbb{E}^N)$,

$$< [M^n], \omega >= \int_{M^n} f_\omega dv_{M^n} .$$

To simplify the notation, if no confusion is possible, we also denote by M^n itself the current $[M^n]$. The boundary of $[M^n]$ is the $(n-1)$-current $[\partial M^n]$ associated to the boundary ∂M^n endowed with the induced orientation.

3. The previous construction can be extended to Lipschitz submanifolds of $\mathbb{E}^N$, i.e., submanifolds which are locally the graph of Lipschitz maps, since such maps are C^1-almost everywhere.

12.2 Rectifiable Currents

Let us now introduce a general kind of currents, associated to *rectifiable* subsets. We deal here with the Hausdorff measure $\mathcal{H}^k$. We refer to [43, pp. 251, 380, 384] and [63, p. 40] for details. A subset A of $\mathbb{E}^N$ is called *k-rectifiable* if it is the image of a bounded subset of $\mathbb{E}^k$ under a Lipschitz map. It can be proved that a k-rectifiable set has a k-dimensional tangent subspace $\mathcal{H}^k$-almost everywhere, from which one deduces that it can be oriented almost everywhere, by assigning a (unit) k-vector U (or a unit k-form) at each tangent vector subspace. There are many possible equivalent definitions of rectifiable currents. Here is the most intuitive one.

Definition 31.

- Let T be a k-current with compact support in $\mathbb{E}^N$. If there exist:

  - A $\mathcal{H}^k$-measurable and k-*rectifiable* subset A of $\mathbb{E}^N$
  - An orientation defined at each point of A admitting a tangent space (consequently $\mathcal{H}^k$-almost everywhere)
  - An integrable function μ with positive integer values defined at each point of A admitting a tangent space, satisfying

$$\int_A \mu d\mathcal{H}^k < \infty$$

  such that

$$\forall \omega \in \mathcal{D}_c^k(\mathbb{E}^N), \, <T, \omega> = \int_A <U, \omega> \mu \mathcal{H}^k,$$

  then T is called the rectifiable k-current associated to the triple (A, U, μ). If no confusion is possible, T will be simply denoted by A.

- The space of rectifiable k-currents is denoted by $\mathcal{R}_k(\mathbb{E}^N)$.
- If T and ∂T are rectifiable, T is called an integral current.
- The space of integral k-currents of $\mathbb{E}^N$ is denoted by $I_k(\mathbb{E}^N)$.

Let us now define the *mass* of a current, which directly generalizes the volume of an oriented submanifold.

Definition 32.

1. The mass of a k-current T of $\mathbb{E}^N$ is the real number $\mathbf{M}(T)$ defined by

$$\mathbf{M}(T) = \sup\{T(\omega) : \omega \in \mathcal{D}_c^k(\mathbb{E}^N) \text{ and } \sup_{m \in \mathbb{E}^N} |\omega_m| \leq 1\}.$$

2. The flat norm of a k-current T of $\mathbb{E}^N$ is the real number $\mathcal{F}(T)$ defined by

$$\mathcal{F}(T) = \inf\{\mathbf{M}(A) + \mathbf{M}(B) : T = A + \partial B, A \in \mathcal{R}_k(\mathbb{E}^N), B \in \mathcal{R}_{k+1}(\mathbb{E}^N)\}.$$

Note that the flat convergence of currents implies the weak convergence of currents. Moreover, the flat norm has an interesting geometrical property: if two "objects" are "close" one to each other, then the difference of their associated currents (when they exist) has a small flat norm. For instance, if C is an oriented curve "close" to the oriented curve C', then $\mathcal{F}(C - C')$ is small (Fig. 12.1).

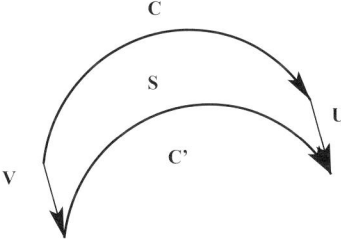

Fig. 12.1 One has $C + U - C' + V = \partial S$. If $\mathbf{M}(U) < \varepsilon$, $\mathbf{M}(V) \leq \varepsilon$, and $\mathbf{M}(S) < \varepsilon^2$, then $\mathcal{F}(C - C') < 3\varepsilon$

12.3 Three Theorems

We mention some classical results on currents, which will be useful in our context. The first theorem gives a characterization of integral currents.

Theorem 35. (The Constancy Theorem) *Let V^k be a (Lipschitz) submanifold of $\mathbb{E}^N$. Let T be a k-integral current of $\mathbb{E}^N$ such that:*

1. $\mathrm{spt}(T) \subset V^k$.
2. $\mathrm{spt}(\partial T) \subset (\partial V^k)$.

Then, there exists $m \in \mathbb{Z}$ such that $T = mV^k$.

The second result is a compactness theorem for integral currents. It provides the means of proving that a sequence of currents converges.

Theorem 36. (The Compactness Theorem) *Let K be a compact subset of $\mathbb{E}^N$ and let c be a positive real number. Let $I_k(K,c)$ be the set of n-integral currents defined on $\mathbb{E}^N$ whose support lies in K and whose mass and boundary mass are bounded by c. Then $I_k(K,c)$ is compact for the flat norm.*

The last result is a homotopy theorem: let

$$f : \mathbb{E}^N \to \mathbb{E}^N$$

and

$$g : \mathbb{E}^N \to \mathbb{E}^N$$

be two smooth maps. A homotopy between f and g is a smooth map

$$h : [0,1] \times \mathbb{E}^N \to \mathbb{E}^N,$$

such that for all $x \in \mathbb{E}^N$

$$h(0,x) = f(x) \text{ and } h(1,x) = g(x).$$

Such a homotopy is said to be *affine* if it satisfies

$$h(t,x) = (1-t)f(x) + tg(x)$$

for all $x \in \mathbb{E}^N$ and for all $t \in [0,1]$.

The following result gives a bound on the mass $\mathbf{M}[h_*([0,1] \times T)$ for every integral current T of $\mathbb{E}^N$ (it can be generalized to any current representable by integration, but we do not use this generalization in this book).

Theorem 37. *Let*

$$f : \mathbb{E}^N \to \mathbb{E}^N \text{ and } g : \mathbb{E}^N \to \mathbb{E}^N$$

be two smooth maps and let h be an affine homotopy between f and g. Let $T \in I_k(\mathbb{E}^N)$. Then

$$\mathbf{M}[h_\sharp([0,1] \times T)] \leq \mathbf{M}(T)(\sup |g - f| \sup\{|Df|^k, |Dg|^k\}).$$

A Final Remark This chapter deals with currents defined on $\mathbb{E}^N$, but the space $\mathbb{E}^N$ can be replaced by any open subset U for more generality. The proof of these theorems can be found in [63, p. 52] or in [43, p. 357] for the Constancy Theorem 35, in [63, p. 64] or [43, p. 414] for the Compactness Theorem 36, and in [43, p. 363] for Theorem 37.

Part V
On Volume

Chapter 13
Approximation of the Volume

We deal here with a basic question: consider a smooth (compact oriented) submanifold M^n of the Euclidean space $\mathbb{E}^N$, $n < N$, and a (measurable) subset W, "close to M^n." Can we approximate the volume $\mathrm{Vol}_n(M^n)$ of M^n by $\mathrm{Vol}_n(W)$? We have seen in Sect. 3.1.3 that the well-known "Lantern of Schwarz" shows that the area of a sequence of triangulations inscribed in a fixed cylinder of $\mathbb{E}^3$ may tend to infinity when the sequence tends to the cylinder for the Hausdorff topology. We give here a general *approximation theorem* by adding a suitable geometric assumption: we assume that the tangent bundle of the sequence tends to the tangent bundle of M^n, in precise sense.

13.1 The General Framework

Let M^n be a smooth compact (embedded) n-dimensional submanifold of $\mathbb{E}^N$. As mentioned in Sect. 11.1, the tubular neighborhood theorem (Theorem 33) asserts that there exists an open subset U_{M^n} of $\mathbb{E}^N$ containing M^n, on which the orthogonal projection pr onto M^n is well defined. In particular, if r is the reach of M^n, then the orthogonal projection pr onto M^n is well defined on $U_r(M^n)$ (Fig. 13.1).

Definition 33. A subset W of $\mathbb{E}^N$ is closely near M^n if it lies in $U_r(M^n)$ (where r is the reach of M^n) and if the restriction of pr to W is one to one (Figs. 13.2 and 13.3).

The mutual behavior of the tangent spaces of both W and M^n will be a crucial geometric invariant in finding approximation and convergence of geometric quantities. That is why we introduce the following result.

Definition 34. Let M^n be a (smooth) submanifold of $\mathbb{E}^N$ and let W be an n-dimensional topological submanifold, C^1-almost everywhere, closely near M^n:

- At every (regular) point m of W, let

$$\alpha_m \in \left[0, \frac{\pi}{2}\right]$$

Fig. 13.1 W is closely near C: every point m of the polygonal line W has a unique orthogonal projection on C

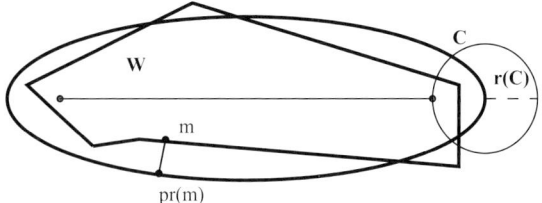

Fig. 13.2 W is not closely near C: although W lies in $U_r(C)$, different points may have the same orthogonal projection on C

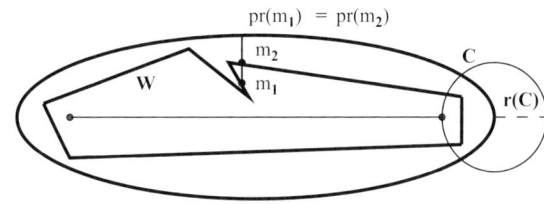

Fig. 13.3 W is not closely near C: although W lies in $U_r(C)$, the restriction to W of the orthogonal projection is not onto

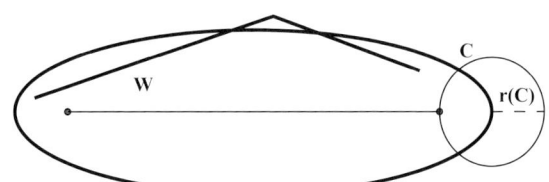

Fig. 13.4 The angular deviation between the polygonal line P and the curve C at m is α_m

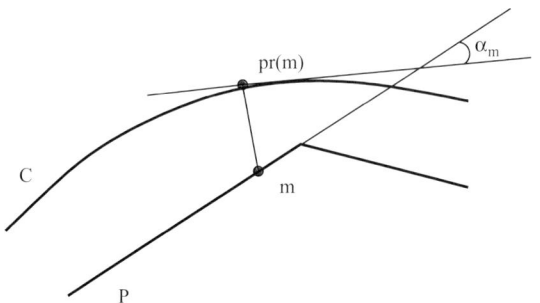

be the angle between the tangent spaces T_mW and $T_{\mathrm{pr}(m)}M^n$.[1] The function α is called the angular deviation function of W with respect to M^n.

• Moreover, we put

$$\begin{cases} \alpha_{\min} & = \inf_{m\in W} \alpha_m, \\ \alpha_{\max} & = \max_{m\in W} \alpha_m. \end{cases} \tag{13.1}$$

• The angle $\alpha_{\max}$ is called the deviation angle of W with respect to M^n (Fig. 13.4).

_____

[1] By definition, the angle between two n-dimensional vector subspaces V_1 and V_2 of $\mathbb{E}^N$ is the absolute value of the determinant of any orthogonal matrix, sending an orthonormal frame of V_1 onto an orthonormal frame of V_2. The angle between two affine n-dimensional subspaces is the angle between their associated vector subspaces.

13.2 A General Evaluation Theorem for the Volume

The following result shows that, if W is an n-dimensional submanifold (with enough regularity to have a tangent space almost everywhere) closely near M^n, then the n-volume of M^n is bounded from above and below by quantities depending on the n-volume of W, the Hausdorff distance between W and M^n, the curvature of M^n, and the deviation angle.

13.2.1 Statement of the Main Result

Theorem 38. *Let M^n be a (compact oriented) C^2-submanifold in $\mathbb{E}^N$. Let W be an n-dimensional submanifold,[2] closely near M^n. Then*

$$\operatorname{Vol}_n(M^n) = \int_W \frac{\cos \alpha_m}{\sum_{k=0}^n \Xi_{k_{\mathrm{pr}(m)}} \delta_m^k} dv_W, \qquad (13.2)$$

where $\Xi_{k_{\mathrm{pr}(m)}}$ denotes the kth-symmetric function associated to the Weingarten endomorphism of M^n in the direction $\xi_{\mathrm{pr}(m)} = \frac{\overrightarrow{\mathrm{pr}(m)m}}{|\mathrm{pr}(m)m|}$ and $\delta_m = |\overrightarrow{\mathrm{pr}(m)m}|$.

13.2.2 Proof of Theorem 38

Strictly speaking, the normal vector field on M^n $\xi_{\mathrm{pr}(m)} = \frac{\overrightarrow{\mathrm{pr}(m)m}}{|\mathrm{pr}(m)m|}$ is only defined when $m \in W$ does not lie in M^n. If the intersection of M^n and W is a (Borel) subset $\mathcal{B}$, then the measure of $\mathcal{B}$ considered as a subset of M^n is obviously the same as the measure of $\mathcal{B}$ considered as a subset of W. Moreover, $\alpha = 0$, $\delta = 0$ on $\mathcal{B}$, and the integrated function in (13.2) equals 1 on $\mathcal{B}$. Consequently, the result is trivial if one restricts the study on $\mathcal{B}$. That is why we suppose that the normal vector field ξ is nowhere null.

The proof of Theorem 38 is a consequence of the study of the differential of the projection map. We summarize its main properties (see [43, 66]).

Lemma 3. *Let M^n be a smooth submanifold of $\mathbb{E}^N$ without boundary and let U_{M^n} be an open subset of $\mathbb{E}^N$ where the map $\mathrm{pr} : U_{M^n} \to M^n$ is well defined. Then:*

1. The map pr is C^1 in U_{M^n}.
2. For all $m \in U_{M^n}$, for all Z orthogonal to $T_{\mathrm{pr}(m)}M^n$, and for all X_m parallel to $T_{\mathrm{pr}(m)}M^n$,

---

[2] As above, with enough regularity to have a tangent space almost everywhere, for instance C^1-almost everywhere.

$$\begin{cases} D\mathrm{pr}(m)(Z_m) &= 0, \\ D\mathrm{pr}(m)(X_m) &= (Id + \delta_m A_{\xi_{\mathrm{pr}(m)}})^{-1}(X_m), \end{cases} \tag{13.3}$$

where $A_{\mathrm{pr}(m)}$ is the Weingarten endomorphism of M^n at the point $\mathrm{pr}(m)$ in the direction $\xi_{\mathrm{pr}(m)} = \dfrac{\overrightarrow{\mathrm{pr}(m)m}}{|\overrightarrow{\mathrm{pr}(m)m}|}$.

3. *In particular, let*

$$(e^1_{\mathrm{pr}(m)}, ..., e^n_{\mathrm{pr}(m)}, v^{n+1}_{\mathrm{pr}(m)}, ..., v^N_{\mathrm{pr}(m)})$$

be a local orthonormal frame of $T_{\mathrm{pr}(m)}\mathbb{E}^N$ such that

$$(e^1_{\mathrm{pr}(m)}, ..., e^n_{\mathrm{pr}(m)})$$

is a unit frame of principal vectors of M^n at $\mathrm{pr}(m)$ in the direction $\xi_{\mathrm{pr}(m)}$, with principal curvatures

$$\lambda^1_{\mathrm{pr}(m)}, ..., \lambda^k_{\mathrm{pr}(m)}, ..., \lambda^n_{\mathrm{pr}(m)},$$

and

$$(v^{n+1}_{\mathrm{pr}(m)}, ..., v^N_{\mathrm{pr}(m)})$$

are normal to M^n. In these two frames, the matrix of

$$D\mathrm{pr}(m) : \mathbb{E}^N \to T_{\mathrm{pr}(m)}M^n$$

is given by

$$\begin{pmatrix} \frac{1}{1+\delta_m\varepsilon_m\lambda^1_{\mathrm{pr}(m)}} & \cdots & 0 & \cdots & 0 & 0 \cdots 0 \\ \cdots & \cdots & \cdots & \cdots & \cdots & \cdots \cdots \cdots \\ 0 & \cdots & \frac{1}{1+\delta_m\varepsilon_m\lambda^k_{\mathrm{pr}(m)}} & \cdots & \cdots & \cdots \cdots 0 \\ \cdots & \cdots & \cdots & \cdots & \cdots & \cdots \cdots \cdots \\ 0 & \cdots & 0 & \cdots & \frac{1}{1+\delta_m\varepsilon_m\lambda^n_{\mathrm{pr}(m)}} & 0 \cdots \cdots \end{pmatrix},$$

where $\delta_m = |\overrightarrow{\mathrm{pr}(m)m}|$ and $\varepsilon_m = \pm 1$ depending on the orientation of the normal local frame.

Sketch of proof of Lemma 3

- The projection map pr from U_{M^n} to M^n is constant on each straight line $[m, \mathrm{pr}(m)]$, from which we deduce the first item.
- Consider now the submanifold V_δ parallel to M^n at distance δ (with δ small enough). The restriction of the projection map pr to V_δ is one to one and its inverse is the one-to-one map

$$p \to p + \delta\xi,$$

where ξ is the suitable unit normal vector field on M^n. Computing the differential of these two one-to-one maps gives the second item.
- The rest of the proof is obvious. $\square$

Consider now the submanifold W lying in U_{M^n}. Since W is differentiable almost everywhere, the function pr is differentiable at almost every point of W. We apply the general coarea formula (Theorem 19):

$$\mathrm{Vol}_n(M^n) = \int_{M^n} dv_{M^n} = \int_W |J_n D\mathrm{pr}(m)| \, dv_W(m),$$

where $J_n D\mathrm{pr}(m)$ is the n-dimensional Jacobian of pr at m. At a point m of W admitting a tangent space, $J_n D\mathrm{pr}(m)$ equals the determinant of the linear map

$$D\mathrm{pr}_{|W} : T_m W \to T_{\mathrm{pr}(m)} M^n.$$

Since W is closely near M^n, $\mathrm{pr}_{|W}$ is one to one between W and M^n. Therefore,

$$\mathrm{Vol}_n(M^n) = \int_M |D\mathrm{pr}_{|W}(m)| \, dv_W(m) = \int_W \frac{\cos \alpha_m}{\sum_{k=0}^n \Xi_{k_{\mathrm{pr}(m)}} \delta_m^k} dv_W.$$

One deduces immediately Theorem 38.

13.3 An Approximation Result

To deduce an approximation result from Theorem 38, we need to introduce a new geometric invariant as follows.

Definition 35. Let M^n be a smooth submanifold of $\mathbb{E}^N$:

- The function

$$\omega_{M^n} : U_{M^n} \to \mathbb{R}$$

  defined by

$$\omega_{M^n}(m) = |A_{\overrightarrow{\mathrm{pr}(m)m}}|_{\mathrm{pr}(m)}$$

  is called the relative curvature function with respect to M^n.
- If W is any subset lying in U_{M^n}, the real number

$$\omega_{M^n}(W) = \sup_{m \in W} \omega_{M^n}(m)$$

  is called the relative curvature of W with respect to M^n.

Remarks

- At every point $m \in U_{M^n}$ outside M^n, one can write

$$\omega_{M^n}(m) = |\overrightarrow{\mathrm{pr}(m)m}||A_\xi|_{\mathrm{pr}(m)},$$

where $\xi_m = \frac{\overrightarrow{\mathrm{pr}(m)m}}{|\overrightarrow{\mathrm{pr}(m)m}|}$. Now, $|A_\xi|_{\mathrm{pr}(m)}$ is the norm of the Weingarten endomorphism of M^n in the direction ξ, i.e., the maximum of the absolute value of the principal curvatures of M^n at $\mathrm{pr}(m)$) in the direction ξ.

• If W lies in $U_r(M^n)$, where r is the reach of M^n, one has

$$\omega_{M^n}(W) \leq 1.$$

In fact, for every $m \in W$, the point $m - 2\overrightarrow{m\mathrm{pr}(m)}$ (i.e., the orthogonal symmetric of m with respect to the tangent space of M^n at $\mathrm{pr}(m)$) belongs to $U_r(M^n)$.

As an immediate consequence of Theorem 38, we get the following result.

Theorem 39. *Let M^n be a (compact oriented) C^2-submanifold in $\mathbb{E}^N$. Let W be an n-dimensional submanifold, closely near M^n. Then*

$$\frac{\cos\alpha_{\max}}{(1 + \omega_{M^n}(W))^n} \mathrm{Vol}_n(W) \leq \mathrm{Vol}_n(M^n) \leq \frac{\cos\alpha_{\min}}{(1 - \omega_{M^n}(W))^n} \mathrm{Vol}_n(W). \qquad (13.4)$$

In particular,

$$|\mathrm{Vol}_n(M^n) - \mathrm{Vol}_n(W)| \leq K(M^n)(\alpha_{\max}^n + \omega_{M^n}(W)) \qquad (13.5)$$

and

$$|\mathrm{Vol}_n(M^n) - \mathrm{Vol}_n(W)| \leq K(M^n)(\alpha_{\max}^n + \mathrm{d}(M^n, W)), \qquad (13.6)$$

where K is a constant depending only on the geometry of M^n.

We deduce the following corollary.

Corollary 2. *Let M^n be a (compact oriented) C^2-submanifold in $\mathbb{E}^N$. Let W_p be a sequence of n-dimensional submanifolds C^1-almost everywhere, closely near M^n. If (Fig. 13.5)*

1. the Hausdorff limit of W_p is M^n when p tends to infinity,

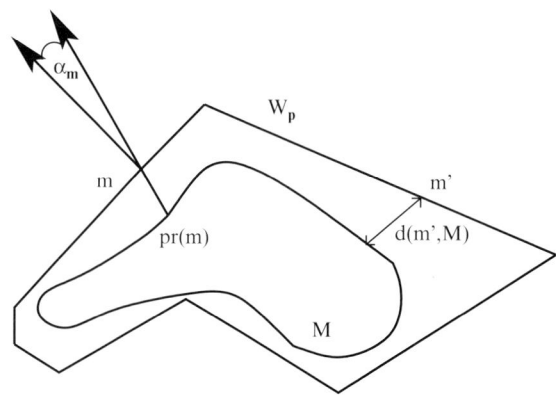

Fig. 13.5 If the Hausdorff distance and the maximum angle between W and M tend to 0, then the volume of W tends to the volume of M

2. *the deviation angle $\alpha_{\max}$ of W_p with respect to M^n tends to 0-almost everywhere when p tends to infinity,*

then

$$\lim_{p \to \infty} \mathrm{Vol}_n(W_p) = \mathrm{Vol}_n(M^n).$$

13.4 A Convergence Theorem for the Volume

As an application of Theorem 38, we deal now with a sequence of polyhedra (which are *PL*-submanifolds and then C^∞ on an open subset of full measure) converging to a smooth submanifold. Following Fu [48], we shall retain the assumption on Hausdorff convergence, but we remove the assumption on the deviation angle and replace it by an assumption on the *fatness* of the polyhedra. Note that this new assumption is *intrinsic*, in the sense that it does not relate the sequence of polyhedra with the smooth submanifold. It only requires that "the angles of the simplices do not tend to 0." We then obtain a nice convergence theorem for the volumes.

13.4.1 The Framework

First of all, we need to improve our background on polyhedra (see Chap. 6). Let P be a polyhedron of $\mathbb{E}^N$, endowed with a fixed triangulation.[3] We denote by:

- $\mathcal{S}_P$ the set of all simplices of P and by $\mathcal{S}_P^k$ the set of all k-simplices of P.
- In particular, we denote by $\mathcal{V}_P$ the set of all vertices of P.
- We denote by $\mathcal{E}_P$ the set of all edges of P.

13.4.1.1 Fatness of a Triangulated Polyhedron

The *fatness* of a polyhedron is the main invariant, which will be involved in our convergence theorem.

Definition 36. Let σ be a k-simplex of $\mathbb{E}^N$ and let P be a polyhedron of $\mathbb{E}^N$:

1. The size of σ is the real number

$$\varepsilon(\sigma) = \max_{e \in \mathcal{E}_\sigma} l(e),$$

where $l(e)$ denotes the length of e.

---

[3] We could consider all the triangulations of P but, to simplify our exposition, we fix one of them.

Fig. 13.6 $\mathcal{A}(t) = \frac{1}{2}l_1 l_2 \sin\alpha$; so, if the quotient of the area of t by the square of the longest edge is "not too small," then each angle of t is "not too small"

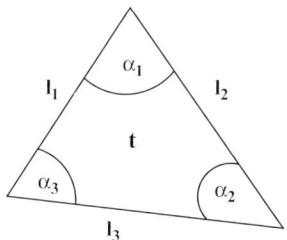

2. The fatness of σ is the real number

$$\Theta(\sigma) = \min_{j \in \{0,\dots,k\}} \left\{ \frac{\mathrm{Vol}_j(\mu)}{\varepsilon(\sigma)^j}, \mu \in \mathcal{S}_\sigma^k \right\}.$$

3. The fatness of P is the real number

$$\Theta(P) = \min_{\sigma \in \mathcal{S}^P} \Theta(\sigma).$$

In other words, "the fatness of a polyhedron is not too small if the angles of its triangles are not too small."

For instance, consider a triangle t in $\mathbb{E}^2$ with e_1, e_1, e_3 its three edges. Suppose that the length of the longest edge is l. Then, each angle α of t satisfies (Fig. 13.6)

$$\frac{\mathcal{A}(t)}{l^2} \le \frac{1}{2} \sin\alpha,$$

where $\mathcal{A}(t)$ denotes the area of t. Consequently,

$$\Theta(t) \le \sin\alpha.$$

13.4.1.2 Polyhedron Closely Inscribed in a Submanifold

Definition 37. Let P be a triangulated polyhedron and M^n be a submanifold of $\mathbb{E}^N$. We say that:

1. P is inscribed in M^n if:

   a. All vertices of P lie in M^n.
   b. All vertices of ∂P lie in ∂M^n.

2. P is closely inscribed in M^n if:

   a. P is inscribed in M^n.
   b. P is closely near M^n.

In particular, if a polyhedron P is closely inscribed in M^n, then P lies in $U_r(M^n)$ and the orthogonal projection onto M^n induces an homeomorphism from P to M^n and from ∂P to ∂M^n. Consequently, its dimension is n.

13.4.2 Statement of the Theorem

The results of this section are due to Fu [48]. Since the smooth submanifolds we consider in this book are C^2 (with or without C^2-boundaries), we state Theorems 40 and 41 with this assumption. However, they can be weakened by considering $C^{1,1}$-submanifolds. The convergence of the volume is the consequence of the following theorem.

Theorem 40. *Let M^n be a (compact) smooth submanifold of $\mathbb{E}^N$. Let $(P_p)_{p \in \mathbb{N}}$ be a sequence of polyhedra closely inscribed in M^n such that:*

1. The Hausdorff limit of P_p tends to M^n when p tends to infinity.
2. The fatness of P_p is uniformly bounded from below by a positive constant:

$$\exists c > 0 \text{ such that } \forall p \geq 0, \Theta(P_p) > c > 0.$$

Then, the sequence of deviation angles $\alpha_{\max_p}$ of $(P_p)_{p \in \mathbb{N}}$ with respect to M^n tends to 0 when p tends to infinity.

The fatness assumption is clearly crucial. We leave the complete proof of this theorem to the reader (see [48] for details). In Chap. 15, we give a precise proof of it in the particular case of two-dimensional polyhedra converging to a smooth surface in $\mathbb{E}^3$. As an immediate consequence of Theorem 40 and Corollary 2, one gets immediately the following result.

Theorem 41. *Let M^n be a smooth submanifold of $\mathbb{E}^N$ with smooth boundary. Let $(P_p)_{p \in \mathbb{N}}$ be a sequence of polyhedra closely inscribed in M^n such that:*

1. The Hausdorff limit of P_p tends to M^n when p tends to infinity.
2. The fatness of P_p is uniformly bounded from below by a positive constant:

$$\exists c > 0 \text{ such that } \forall p \geq 0, \Theta(P_p) > c > 0.$$

Then,

$$\lim_{p \to \infty} \mathrm{Vol}_n(P_p) = \mathrm{Vol}_n(M^n).$$

Chapter 14
Approximation of the Length of Curves

We have seen in Chap. 13 that the length of a curve is classically defined as the supremum of the lengths of polygonal lines inscribed in it. Our purpose here is to compare the length of a given smooth curve with the length of a curve close to it, or more precisely with the length of a polygonal line inscribed in it.

14.1 A General Approximation Result

Adapting our general Theorem 38, we get immediately the following result.

Theorem 42. *Let γ be a (compact) C^2-curve in $\mathbb{E}^N$ and C be a curve (differentiable almost everywhere), closely near S. Then,*

$$l(\gamma) = \int_C \frac{\cos \alpha_m}{1 + k_{\mathrm{pr}(m)} < n, \overrightarrow{\mathrm{pr}(m)m} >} dm, \tag{14.1}$$

where:

- *For (almost) every point $m \in C$, α is the deviation angle function.*
- *$k_{\mathrm{pr}(m)}$ is the curvature of γ at $\mathrm{pr}(m)$.*
- *n denotes the principal normal vector of γ.*

In other words, one obtains the length of γ by integrating *on* C a combination of the distance between γ and C, the deviation angle function, and the curvature function of γ.

Let U_γ be a neighborhood of γ on which the orthogonal projection onto γ is well defined. Introducing the relative curvature defined in Definition 35 and the notations (13.1), we get from Theorem 42 the following result.

Theorem 43. *If γ is a (compact) C^2-curve in $\mathbb{E}^N$ and C is a curve (differentiable almost everywhere), closely near S, then*

$$\frac{\cos\alpha_{\max}}{1+\omega_\gamma(C)}l(C) \le l(\gamma) \le \frac{\cos\alpha_{\min}}{1-\omega_\gamma(C)}l(C). \tag{14.2}$$

14.2 An Approximation by a Polygonal Line

Theorem 43 can be improved if the approximating curve is a polygonal line closely inscribed in γ. This is the goal of this section.

Proposition 8. *Let γ be a compact regular smooth curve of $\mathbb{E}^N$ with end points p and q. Assume that $[p,q] \subset U_\gamma$. Then:*

1. The length $l(\gamma)$ of γ satisfies

$$|\overrightarrow{pq}| \le l(\gamma) \le \frac{|\overrightarrow{pq}|}{1-\omega_\gamma(pq)}. \tag{14.3}$$

2. The angle $\theta_p \in [0,\frac{\pi}{2}]$ between $\overrightarrow{pq}$ and the tangent vector t_p of γ at p satisfies (Fig. 14.1)

$$\sin\theta_p \le \frac{k_{\gamma_{\max}}l(\gamma)}{2}, \tag{14.4}$$

where $k_{\gamma_{\max}}$ is the maximum of the curvature of γ.

In particular, we deduce immediately from (14.3) the following result.

Corollary 3. *Let γ be a smooth curve embedded in $\mathbb{E}^N$ and P be a polygonal line closely inscribed in it. Then*

$$l(P) \le l(\gamma) \le \frac{1}{1-\omega_\gamma(P)}l(P).$$

Proof of Proposition 8

1. The double inequality (14.3) is a direct consequence of (14.2).
2. Without loss of generality, let us assume that

$$\gamma:[0,l] \to \mathbb{E}^N$$

is an arc length parametrization of the curve. Then

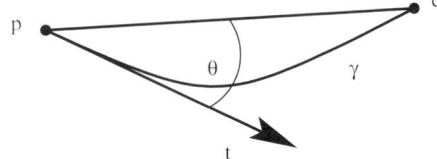

Fig. 14.1 θ is the angle between the tangent vector at p and the chord pq

$$\gamma(0) = p \text{ and } \gamma(l) = q.$$

Using a Taylor–Lagrange expansion of γ and setting $l = l(\gamma)$, we obtain

$$\gamma(l) - \gamma(0) = l\gamma'(0) + \int_0^l (l-s)\gamma''(s)ds.$$

Thus,

$$t_p = \gamma'(0) = \frac{\overrightarrow{pq}}{l} - \frac{1}{l}\int_0^l (l-s)\gamma''(s)ds.$$

Since $|\gamma''| = k$,

$$\left|\frac{1}{l}\int_0^l (l-s)\gamma''(s)ds\right| \le \frac{k_{\gamma\text{max}}l(\gamma)}{2},$$

from which we deduce Proposition 14.4. $\quad\square$

To end this section, let us mention the result of Cohen-Steiner and Edelsbrunner, which uses Morse theory to give a bound on the difference of the lengths of two curves, in terms of their total curvatures and their *Fréchet distance* [45]. The interest of this result is that it does not use the angle between the tangent vector fields of the curves, replacing the Hausdorff distance by the Fréchet one.

Theorem 44. *Let γ_1 and γ_2 be two smooth curves in $\mathbb{E}^N$. Then:*

• *If γ_1 and γ_2 are closed,*

$$|l(\gamma_1) - l(\gamma_2)| \le \frac{2s_{N-1}}{s_N}\left[\int_{\gamma_1} k_{\gamma_1}ds + \int_{\gamma_2} k_{\gamma_2}ds - 2\pi\right]d_F(\gamma_1, \gamma_2).$$

• *If γ_1 and γ_2 are not closed,*

$$|l(\gamma_1) - l(\gamma_2)| \le \frac{2s_{N-1}}{s_N}\left[\int_{\gamma_1} k_{\gamma_1}ds + \int_{\gamma_2} k_{\gamma_2}ds + \pi\right]d_F(\gamma_1, \gamma_2),$$

where $d_F(\gamma_1, \gamma_2)$ denotes the Fréchet distance between γ_1 and γ_2.

The details of the proof can be found in [33].

Chapter 15
Approximation of the Area of Surfaces

15.1 A General Approximation of the Area

Let S be a (smooth compact) C^2-surface in $\mathbb{E}^3$ and W be a surface closely near S. Then, Theorem 38 can be stated as follows:

$$A(S) = \int_W \frac{\cos \alpha_m}{1 + \delta_m H_{\mathrm{pr}(m)} + \delta_m^2 G_{\mathrm{pr}(m)}} dv_W(m), \qquad (15.1)$$

where α is the angular deviation function of W with respect to S (introduced in Definition 34), H is the mean curvature of S, and G is the Gauss curvature of S.

Let ω_S be the relative curvature function on S (introduced in Definition 35). We deduce from Theorem 39 the following result [65], [66].

Corollary 4. *Let S be a (smooth compact) surface and W be a surface closely near S, differentiable almost everywhere:*

1. Then

$$\frac{\cos \alpha_{\max}}{(1 + \omega_S(W))^2} A(W) \le A(S) \le \frac{\cos \alpha_{\min}}{(1 - \omega_S(W))^2} A(W).$$

2. In particular,

$$|A(S) - A(W)| \le K(S)(\alpha_{\max}^2 + \omega_S(W))$$

and

$$|A(S) - A(W)| \le K(S)(\alpha_{\max}^2 + d(S,W)),$$

where $K(S)$ is a constant depending on S and $d(S,W)$ denotes the Hausdorff distance between S and W.

In particular, we have the following corollary.

Corollary 5. *Let S be a (smooth compact) surface and* $(W_p)_{p \in \mathbb{N}}$ *be a sequence of surfaces of* $\mathbb{E}^3$*, such that:*

1. The Hausdorff distances $\mathrm{d}(W_p, S)$ *tend to 0 when p tends to* $+\infty$*.*
2. The angular deviations $\alpha_{\max_p}$ *tend to 0 when p tends to* $+\infty$*.*

Then

$$\lim_{p \to \infty} \mathcal{A}(W_p) = \mathcal{A}(S).$$

15.2 Triangulations

We have seen in Chap. 13 that a "good" approximation of the area of a (smooth) surface is given by the area of a close surface if the tangent spaces of both surfaces are close (Theorem 39). We now deal with the particular case of a *triangulation T closely inscribed* in a surface S of $\mathbb{E}^3$. We shall show that we can control the angular deviation $\alpha_{\max}$ of T by the *shape* of the triangles of T, improving the convergence Theorem 40.

Before dealing with approximation by triangulations, let us improve our knowledge on the basic geometric invariants defined on a two-dimensional triangulation T of $\mathbb{E}^3$. We denote by $\mathcal{T}_T$ the set of triangles of T and by t a generic triangle of T.

15.2.1 Geometric Invariant Associated to a Triangle

First of all, we take the notations of Definition 36:

- The length of the longest edge of t is denoted by $\varepsilon(t)$.
- $\mathcal{A}(t)$ denotes the area of t.
- The *circumradius* of t is denoted by $r(t)$.
- The *fatness* of t is the real number

$$\Theta(t) = \frac{\mathcal{A}(t)}{\varepsilon(t)^2}.$$

Let us now introduce a new geometric invariant adapted to our purpose.

Definition 38. The *rightness* of a triangle t is the real number

$$\mathrm{rig}(t) = \sup_{v \in \mathcal{V}t} |\sin \angle(v)|,$$

where $\angle(v)$ is the angle at v of t.

15.2.2 Geometric Invariant Associated to a Triangulation

Globally:

- As usual, the *area* $\mathcal{A}(T)$ is the sum of the areas of all the triangles of T.
- The *circumradius* $r(T)$ of T is

$$\max_{t \in \mathcal{T}_T} r(t).$$

- The *height* of T is

$$\varepsilon(T) = \max_{t \in \mathcal{T}_T} \varepsilon(t).$$

- As in Definition 36, the *fatness* of T is

$$\Theta(T) = \min_{t \in \mathcal{T}_T} \Theta(t).$$

- The *rightness* of T is

$$\mathrm{rig}\ (T) = \min_{t \in \mathcal{T}_T} \mathrm{rig}\ (t).$$

15.3 Relative Height of a Triangulation Inscribed in a Surface

In this section, we assume that a triangulation T is *closely near* a smooth surface S in the sense of Definition 33. Our main result will be proved in the more restrictive situation where the triangulation is closely *inscribed* in S in the sense of Definition 37. After the definition of the *relative curvature* in Chap. 13, which is a *mixed* invariant depending on both the geometry of S and a surface close to it, we define now another *mixed* invariant associated S and T.

Definition 39. The relative height of T with respect to S is the real number defined by

$$\pi_S(T) = \sup_{t \in T} \sup_{m \in t} \varepsilon(T) |h|_{\mathrm{pr}(m)},$$

where $|h|_{\mathrm{pr}(m)}$ denotes as usual the norm of the second fundamental form of S at $\mathrm{pr}(m)$.[1]

The following simple geometric result compares the relative height $\pi_S(T)$ with the relative curvature $\omega_S(T)$ defined in Chap. 13.

Proposition 9. *Let T be a triangulation closely inscribed in a smooth surface S. Then:*

1. The Hausdorff distance δ_t between any triangle t and its projection $\mathrm{pr}(t)$ onto S satisfies

---

[1] Or equivalently the maximum of the absolute value of the two principal curvatures of S at $\mathrm{pr}(m)$.

$$\delta_t \le \varepsilon(t) \le 2r(t).$$

2. *Moreover,* $\omega_S(t) \le \pi_S(t)$.

Proof of Proposition 9

1. Let p be a point of a triangle t of T. Since T is inscribed in S,

$$d(p, \mathrm{pr}(t)) \le d(p,v) \le \varepsilon(t),$$

where v denotes any vertex of t. Conversely, if m is a point of $\mathrm{pr}(t)$, consider the intersection point p of the normal line to S at m with t. Since T is closely inscribed in S, m is the (unique) orthogonal projection of p onto S and

$$d(m,t) = d(m,p) \le d(p,v) \le \varepsilon.$$

Consequently, $\delta_t \le \varepsilon(t)$.

2. The second item is an immediate consequence of (1). $\qquad\square$

15.4 A Bound on the Deviation Angle

We study here the *angular deviation* $\alpha_{\max}$ between the normal of a smooth surface S and the normal of a triangulation T closely inscribed in it.

15.4.1 Statement of the Result and Its Consequences

Theorem 45. *Let S be a smooth surface and T be a triangulation closely inscribed in S. Then, the deviation angle $\alpha_{\max}$ between S and T satisfies*

$$\sin \alpha_{\max} \le \left(\frac{\sqrt{10}}{2\,\mathrm{rig}(T)\,(1 - \omega_S(T))} + \frac{1}{1 - \omega_S(T)} \right) \pi_S(T).$$

The proof of this theorem will be given at the end of this section. Let us give some of its consequences.

Corollary 6. *Let S be a smooth surface and T be a triangulation closely inscribed in S. If*

$$\pi_S(T) \le \frac{1}{2},$$

then the angular deviation $\alpha_{\max}$ of T with respect to S satisfies

$$\sin \alpha_{\max} \le \left(\frac{4}{\mathrm{rig}(T)} + 2 \right) \pi_S(T).$$

As a consequence, we obtain the following convergence result.

Corollary 7. *Let S be a (compact orientable) smooth surface in $\mathbb{E}^3$. Let T_n be a sequence of triangulations closely inscribed in S. If*

1. *the length of the edges of T_n tends to zero when n tends to infinity,*
2. *the rightness of T_n is (uniformly) bounded from below by a positive constant,*

then the sequence of angular deviations of T_n with respect to S tends to 0 when n tends to infinity.

Since S is compact, the second condition may be weakened by asking that the sequence $\pi_S(T_n)$ tends to zero when n tends to infinity (in some sense, the length of the edges may be "large" when the curvature is "small").

The geometric quantities $\pi_S(T)$ and $\mathrm{rig}\,(T)$ are linked to the circumradii of the triangles of T. If t is a triangle of T, then we have

$$|h|_{\mathrm{pr}(t)}r(t) = \frac{|h|_{\mathrm{pr}(t)}\varepsilon(t)}{2\mathrm{rig}(t)} = \frac{\pi_t(\mathrm{pr}(t))}{2\mathrm{rig}(t)}. \tag{15.2}$$

This implies the following result.

Corollary 8. *Let S be a smooth surface and T be a triangulation closely inscribed in S. Then*

$$\alpha_{\max} \leq \mathbf{C}(S)r(T),$$

where $\mathbf{C}(S)$ is a constant depending on S. In particular,

$$\alpha_{\max} = O(r(T)).$$

15.4.2 Proof of Theorem 45

The proof of Theorem 45 needs technical preparatory lemmas and propositions.

15.4.2.1 A Purely Geometric Result

Lemma 4. *Let t be a triangle whose vertices are p, p_1, and p_2. If $\alpha_p \in [0, \frac{\pi}{2}\mathrm{rig(t)}]$ denotes the angle between a normal to the triangle and the axis (O,z), then:*

1.

$$cos^2(\alpha_p) = \frac{\cos^2\theta_1\cos^2\theta_2 - \sin^2\theta_1\sin^2\theta_2 - \cos^2\gamma + 2\cos\gamma\sin\theta_1\sin\theta_2}{\sin^2(\theta_2-\theta_1) + 2\sin\theta_1\sin\theta_2\cos(\theta_2-\theta_1) + \cos^2\theta_1\cos^2\theta_2 - \sin^2\theta_1\sin^2\theta_2 - \cos^2\gamma},$$

where $\theta_i \in \left[-\frac{\pi}{2}, \frac{\pi}{2}\right]$ is the angle between $p\vec{p}_i$ and the orthogonal projection of $p\vec{p}_i$ onto the plane orthogonal to (O,z) which contains p ($\theta_i \geq 0$ if and only if the third component of $p\vec{p}_i$ is positive) and $\gamma \in]0, \pi[$ is the angle of t at p.

2. In particular, if $|\sin\theta_1| \leq \varepsilon$ *and* $|\sin\theta_2| \leq \varepsilon$, *then*

$$\sin(\alpha_p) \leq \frac{\sqrt{10}\varepsilon}{\sin\gamma}.$$

Proof of Lemma 4 It is a simple computation, which we do not reproduce here. The complete proof can be found in [66]. □

15.4.2.2 Comparing the Length of a Geodesic and Its Chord

Proposition 10. *Let S be a smooth compact surface of* $\mathbb{E}^3$, U_S *be a neighborhood of S where the map* $\mathrm{pr}: U_S \to S$ *is well defined, and p and q be two points on S such that* $[p,q] \subset U_S$ *and* $\mathrm{pr}(]p,q[) \subset S \setminus \partial S$. *Then, the distance* l_{pq} *between p and q on S satisfies*

$$\overrightarrow{pq} \leq l_{pq} \leq \frac{1}{1-\omega_S(pq)}\overrightarrow{pq}.$$

Proof of Proposition 10 The left inequality is trivial. On the other hand, since $\mathrm{pr}([p,q])$ is a curve on S, its length is larger than the length l_{pq} of the minimal geodesic on S whose ends are p and q. Therefore, using the mean value theorem, one has

$$l_{pq} \leq l(\mathrm{pr}([p,q])) \leq \sup_{m \in]p,q[} |D\,\mathrm{pr}(m)|\,pq.$$

Since Proposition 3 implies that

$$|D\,\mathrm{pr}(m)| \leq \frac{1}{1-|\mathrm{pr}(m)-m||h|_{\mathrm{pr}(m)}} \leq \frac{1}{1-\omega_S(pq)},$$

Proposition 10 is proved. □

15.4.2.3 Comparing the Normals at a Vertex

Proposition 11. *Let S be a smooth surface, t be a triangle closely inscribed in S, and p be a vertex of t. Then, the angle* $\alpha_p \in \left[0,\frac{\pi}{2}\right]$ *between the normals of S and t at p satisfies*

$$\sin(\alpha_p) \leq \frac{\sqrt{10}\,\pi_S(t)}{2\,\sin\gamma_p\,(1-\omega_S(t))},$$

where γ_p *is the angle of t at p.*

This proposition is a consequence of the following lemma.

Lemma 5. *Let S be a smooth surface and let* $p,q \in S$ *such that* $[p,q] \subset S$. *Then, the angle* $\theta \in \left[0,\frac{\pi}{2}\right]$ *between* $\overrightarrow{pq}$ *and the orthogonal projection of* $\overrightarrow{pq}$ *onto* T_pS *satisfies*

$$\sin\theta \le \frac{|h_S| l_{pq}}{2},$$

where $|h_S|$ denotes the supremum over S of the norm of the second fundamental form of S and l is the distance on S between p and q.

The proof of Lemma 5 is similar to that of Proposition 8.

Proof of Proposition 11 Denote by l_1 the distance on S between p and p_1, and by l_2 the distance on S between p and p_2. Since T is closely inscribed in S, thanks to Corollary 10, we obtain

$$l_1 \le \frac{pp_1}{1 - \omega_S(T)} \le \frac{\varepsilon_t}{1 - \omega_S(t)} \quad \text{and} \quad l_2 \le \frac{\varepsilon_t}{1 - \omega_S(t)}.$$

Therefore, Lemma 5 implies that

$$\sin\theta_1 \le \frac{|h|\, \mathrm{pr}_{(t)} l_1}{2} \le \frac{\pi_S(t)}{2(1 - \omega_S(t))} \quad \text{and} \quad \sin\theta_2 \le \frac{\pi_S(t)}{2(1 - \omega_S(t))}.$$

Then, Lemma 4 implies that

$$\sin(\alpha_p) \le \frac{\sqrt{10}}{\sin\gamma_p} \frac{\pi_S(t)}{2(1 - \omega_S(t))} = \frac{\sqrt{10}\,\pi_S(t)}{2\,\sin\gamma_p\,(1 - \omega_S(t))}. \qquad \square$$

15.4.2.4 Comparing the Normals of a Smooth Surface

Proposition 12. *Let S be a smooth compact oriented surface of $\mathbb{E}^3$, t be a triangle closely inscribed in S, and p and s be two points on T. Then, the angle $\alpha_{sp} \in \left[0, \frac{\pi}{2}\right]$ between two normals $\xi_{\mathrm{pr}(p)}$ and $\xi_{\mathrm{pr}(s)}$ at $\mathrm{pr}(p)$ and $\mathrm{pr}(s)$ satisfies*

$$\sin(\alpha_{sp}) \le \frac{\pi_S(t)}{1 - \omega_{\mathrm{pr}(t)}(t)}.$$

This proposition is the consequence of the following lemma, which is a direct application of the mean value theorem.

Lemma 6. *Let S be a smooth compact oriented surface of $\mathbb{E}^3$ and a and b be two points of S. Then, the angle $\alpha_{ab} \in \left[0, \frac{\pi}{2}\right]$ between two normals ξ_a and ξ_b at a and b satisfies*

$$\sin(\alpha_{ab}) \le |h_S| l_{ab},$$

where l_{ab} is the distance on S between a and b.[2]

---

[2] By definition, l_{ab} is the infimum of the lengths of the curves on S linking a and b.

The proof of Proposition 12 is an obvious consequence of Lemma 6 and Corollary 4.

Theorem 45 can be immediately deduced from Propositions 11 and 12, since

$$\sin(\alpha_s) \le \sin(\alpha_p) + \sin(\alpha_{sp}).$$

15.5 Approximation of the Area of a Smooth Surface by the Area of a Triangulation

A direct application of Theorem 45 gives the following result.

Corollary 9. *Let S be a (compact orientable) C^2-surface in $\mathbb{E}^3$ and T be a triangulation closely inscribed in S. If*

$$\left(\frac{4}{\mathrm{rig}(T)} + 2 \right) \pi_S(T) \le 1,$$

then the area of S satisfies

$$\frac{\sqrt{1 - (\frac{4}{\mathrm{rig}(T)} + 2)^2 \pi_S(T)^2}}{(1 + \omega_S(T))^2} \mathcal{A}(T) \le \mathcal{A}(S) \le \frac{1}{(1 - \omega_S(T))^2} \mathcal{A}(T).$$

Corollary 10. *Let S be a (compact orientable) C^2-surface in $\mathbb{E}^3$. Let T_n be a sequence of triangulations closely inscribed in S. If*

1. *the length of the edges of T_n tends to zero when n tends to infinity,*
2. *the rightness of T_n is (uniformly) bounded from below by a positive constant,*

then

$$\lim_{n \to \infty} \mathcal{A}(T_n) = \mathcal{A}(S).$$

Using (15.2), we also have the following result.

Corollary 11. *Let S be a smooth surface and T be a triangulation closely inscribed in S. Then*

$$|\mathcal{A}(T) - \mathcal{A}(S)| \le \mathbf{C}(S) r(T),$$

where $\mathbf{C}(S)$ is a constant depending on S.

Part VI
The Steiner Formula

Chapter 16
The Steiner Formula for Convex Subsets

This chapter is devoted to the computation of the volume of the parallel body of a convex body $\mathcal{K}$ at distance ε (see the definition below). It appears that the convexity of $\mathcal{K}$ implies that this volume is polynomial in ε, the coefficients $(\Phi_k(\mathcal{K}), 0 \leq k \leq N)$ depending on the geometry of $\mathcal{K}$ [77]. Up to a constant, these coefficients (called the *Quermassintegrale* of Minkowski) are the valuations, which appear in Definition 23 and Theorem 28 of Hadwiger. Moreover, these coefficients can be easily evaluated when the boundary of $\mathcal{K}$ is smooth: up to a constant depending on N, they are the integral of the kth-*mean curvatures of the boundary* $\partial \mathcal{K}$ of $\mathcal{K}$. That is why they are good candidates to generalize the curvatures of a smooth hypersurface: they can be defined for any convex subset, even if its boundary is not of class C^2. We shall say that the sequence $\Phi_k(\mathcal{K})$ defines the kth-*mean curvatures* of $\mathcal{K}$ (and by extension if there is no possible confusion, the kth-*mean curvatures* of $\partial \mathcal{K}$). Of course, the explicit evaluation of these curvatures cannot be done by differentiations of a parametrization of the boundary, because of the lack of differentiability. We shall directly evaluate them for convex polyhedra. All these techniques will be generalized in the next chapters to objects which are not convex, but which have geometrical properties close to those of convex bodies.

16.1 The Steiner Formula for Convex Bodies (1840)

The *Steiner formula* gives the behavior of the volume of the tube $\mathcal{K}_\varepsilon$ of radius ε of a convex subset $\mathcal{K}$ of $\mathbb{E}^N$ (see Definition 9). It will be proved that it is a polynomial in ε, whose coefficients depend only on $\mathcal{K}$ and the dimension N of the ambient space.

Theorem 46. *Let $\mathcal{K}$ be a convex subset of $\mathbb{E}^N$. Then,*

$$\mathrm{Vol}_N(\mathcal{K}_\varepsilon) = \sum_{k=0}^{N} \Phi_k(\mathcal{K}) \varepsilon^k, \forall \varepsilon \geq 0. \tag{16.1}$$

Moreover,

$$\Phi_0(\mathcal{K}) = \mathrm{Vol}_N(\mathcal{K}). \tag{16.2}$$

The coefficients $\Phi_k(\mathcal{K})$ are called the *Quermassintegrale* of Minkowski and the polynomial itself is called the *Steiner polynomial*.

Proof of Theorem 46 The following proof uses the Cauchy formula, given in Theorem 27. We proceed by induction on the dimension N of the ambient space:

- For $N = 1$, it is trivial, since in this case $\mathcal{K}$ is a point or segment, and

$$\mathrm{Vol}_1(\mathcal{K}_\varepsilon) = \mathrm{Vol}_1(\mathcal{K}) + 2\varepsilon.$$

- Suppose that the formula is true up to $N-1$. Let ρ be a real number such that $0 \le \rho \le \varepsilon$. Consider the projection $\mathcal{K}_{\rho_{P^{N-1}}}$ of $\mathcal{K}_\rho$ onto any hyperplane P^{N-1} of $\mathbb{E}^N$. The subset $\mathcal{K}_{\rho_{P^{N-1}}}$ is still convex and one can apply the induction assumption

$$\mathrm{Vol}_{N-1}(\mathcal{K}_{\rho P^{N-1}}) = \sum_{k=0}^{N-1} \Phi_k(\mathcal{K}_{P^{N-1}})\rho^k, \forall \rho \ge 0. \tag{16.3}$$

- Using an integration by slice (Theorem 32), one has

$$\mathrm{Vol}_N(\mathcal{K}_\varepsilon) = \mathrm{Vol}_N(\mathcal{K}) + \int_0^\varepsilon \mathrm{Vol}_{N-1}(\partial \mathcal{K}_\rho)d\rho.$$

- Now, we evaluate $\mathrm{Vol}_N(\partial \mathcal{K}_\rho)$ by the Cauchy formula (Theorem 27)

$$\mathrm{Vol}_N(\mathcal{K}_\varepsilon) = \mathrm{Vol}_N(\mathcal{K}) + \int_0^\varepsilon c(N) \int_{G(N,N-1)} \mathrm{Vol}_{N-1}(\mathcal{K}_{\rho_{P^{N-1}}})dP^{N-1}d\rho.$$

- We replace $\mathrm{Vol}_{N-1}(\mathcal{K}_{\rho_{P^{N-1}}})$ by its value in (16.3)

$$\mathrm{Vol}_N(\mathcal{K}_\varepsilon) = \mathrm{Vol}_N(\mathcal{K}) + \int_0^\varepsilon c(N) \int_{G(N,N-1)} \left(\sum_{k=0}^{N-1} \Phi_k(\mathcal{K}_{P^{N-1}})\rho^k\right)dP^{N-1}d\rho,$$

and integrating in ρ, we find that

$$\mathrm{Vol}_N(\mathcal{K}_\varepsilon) = \sum_{k=0}^N \Phi_k(\mathcal{K})\varepsilon^k \tag{16.4}$$

with suitable Φ_k. $\square$

An Important Remark Although we do not specify the dimension of the ambient space in the notations of the *Quermassintegrale*, their values depend on it. In fact, the parallel body of $\mathcal{K}$ considered in $\mathbb{E}^N$ is different to the one considered in $\mathbb{E}^{N'}, N \ne N'$. To avoid any confusion, it is better to denote the *Quermassintegrale* by $\Phi_{k,N}(\mathcal{K})$. If we consider the standard embedding of $\mathbb{E}^N$ in $\mathbb{E}^{N+1}$, then

$$\frac{1}{b_k}\Phi_{k,N}(\mathcal{K}) = \frac{1}{b_{k+1}}\Phi_{k+1,N+1}(\mathcal{K}), \tag{16.5}$$

where b_k denotes the volume of the unit k-dimensional ball. In other words, introducing the quantities

$$\Lambda_{k,N}(\mathcal{K}) = \frac{1}{b_{N-k}}\Phi_{N-k,N}(\mathcal{K}), \tag{16.6}$$

(16.4) can be written

$$\mathrm{Vol}_N(\mathcal{K}_\varepsilon) = \sum_{k=0}^{N}\Lambda_{N-k,N}(\mathcal{K})b_k\varepsilon^k, \forall \varepsilon \geq 0 \tag{16.7}$$

and (16.5) becomes

$$\Lambda_{i,N} = \Lambda_{i,N+1}. \tag{16.8}$$

We conclude that the coefficients $\Lambda_{i,N}$ depend only on (i and) the geometry of $\mathcal{K}$ and not on the dimension of the ambient space. One says that they are *intrinsic*. Consequently, since any convex subset $\mathcal{K}$ is a convex body of an affine space of minimum dimension P, the only interesting *Quermassintegrale* are the $\Phi_{k,P}(\mathcal{K})$. So, without loss of generality and when no confusion is possible, we can assume from now on that $\mathcal{K}$ is a convex body of $\mathbb{E}^N$.

Warning! The word *intrinsic* has different meanings in geometry. Here, it means that the studied quantity does not depend on the dimension of the ambient space.

16.2 Examples: Segments, Discs, and Balls

The *Steiner formula* can be checked in particular cases by a direct computation: as we have seen in the proof of Theorem 46, the simplest (trivial) example occurs with $N = 1$. In this case, $\mathcal{K}$ is an interval I and we have

$$l(I_\varepsilon) = l(I) + 2\varepsilon,$$

where l denotes the length (Fig. 16.1).

When $N = 2$ or 3, we have the following results, in which A denotes the area:

1. If $N = 2$ and

   - p is a point in $\mathbb{E}^2$, then (Fig. 16.2)

$$A(\{p\}_\varepsilon) = \pi\varepsilon^2;$$

Fig. 16.1 The tube of radius ε of a segment I in $\mathbb{R}$

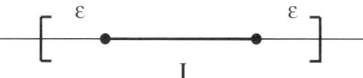

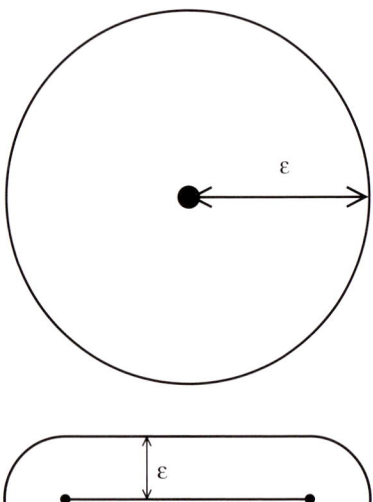

Fig. 16.4 The tube of radius
ε of a disc in $\mathbb{E}^2$

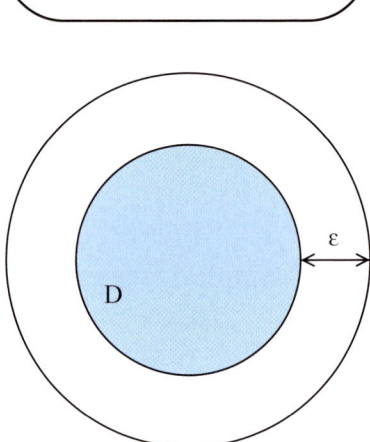

- I is a segment in $\mathbb{E}^2$, then (Fig. 16.3)

$$A(I_\varepsilon) = 2l(I)\varepsilon + \pi\varepsilon^2;$$

- if D is a disc in $\mathbb{E}^2$, then (Fig. 16.4)

$$A(D_\varepsilon) = A(D) + l(\partial D)\varepsilon + \pi\varepsilon^2.$$

2. If $N = 3$ and

- p is a point in $\mathbb{E}^3$, then (Fig. 16.5)

$$\text{Vol}_3(\{p\}_\varepsilon) = \frac{4}{3}\pi\varepsilon^3;$$

Fig. 16.5 The tube of radius
ε of a point in $\mathbb{E}^3$

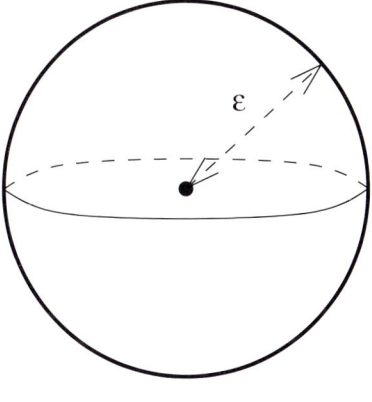

Fig. 16.6 The tube of radius
ε of a segment in $\mathbb{E}^3$

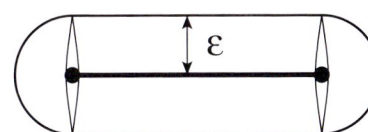

Fig. 16.7 The tube of radius
ε of a ball of radius r in $\mathbb{E}^3$

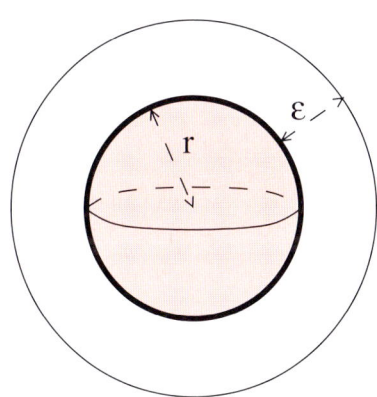

- I is a segment in $\mathbb{E}^3$, then (Fig. 16.6)

$$\mathrm{Vol}_3(I_\varepsilon) = \pi l(I)\varepsilon^2 + \frac{4}{3}\pi\varepsilon^3;$$

- if B is a ball of radius r in $\mathbb{E}^3$, then

$$\mathrm{Vol}_3(B_\varepsilon) = \mathrm{Vol}_3(B) + 4\pi r^2\varepsilon + 4\pi r\varepsilon^2 + \frac{4}{3}\pi\varepsilon^3,$$

where r denotes the radius of the ball B (Fig. 16.7).

16.3 Convex Bodies in $\mathbb{E}^N$ Whose Boundary is a Polyhedron

In this section, we deal with the case where the boundary $\partial \mathcal{K}_\varepsilon$ of the convex body $\mathcal{K}$ is a polyhedron (we can reduce our study to the case of a body using (16.8)). Let us compute the coefficients of the *Steiner polynomial*. We use the notations and definitions given in Chap. 6.

Theorem 47. *Let $\mathcal{K}$ be a convex body of $\mathbb{E}^N$ whose boundary $\partial \mathcal{K}$ is a polyhedron. Then*

$$\mathrm{Vol}_N(\mathcal{K}_\varepsilon) = \sum_{k=0}^{k=N} \Phi_k(\mathcal{K})\varepsilon^k. \tag{16.9}$$

The coefficient $\Phi_k(\mathcal{K})$ can be evaluated as follows:

- *If σ^N is any simplex of $\mathbb{E}^N$, then*

$$\Phi_k(\sigma^N) = \sum_{\sigma^{N-k} \subset \partial \sigma^N} \mathrm{Vol}_{N-k}(\sigma^{N-k})(\sigma^{N-k}, \sigma^N)^*. \tag{16.10}$$

- *More generally, for any convex body $\mathcal{K}$,*

$$\begin{aligned}
\Phi_0(\mathcal{K}) &= \mathrm{Vol}_N(\mathcal{K}), \\
\Phi_1(\mathcal{K}) &= \mathrm{Vol}_{N-1}(\partial \mathcal{K}), \\
\Phi_N(\mathcal{K}) &= \mathrm{Vol}_N(\mathbb{B}^N),
\end{aligned} \tag{16.11}$$

and $\forall k, 1 < k < N$,

$$\Phi_k(\mathcal{K}) = \sum_{\sigma^{N-k} \subset \partial K} \mathrm{Vol}_{N-k}(\sigma^{N-k})(\sigma^{N-k}, \sigma^N)^*, \tag{16.12}$$

where:

- *σ^{N-k} denotes a generic $(N-k)$-face of $\partial \mathcal{K}$.*
- *$(\sigma^{N-k}, \sigma^N)^*$ denotes the normalized dihedral external angle[1] of σ^{N-k} in σ^N.*
- *$\mathbb{B}^N$ denotes the unit ball of $\mathbb{E}^N$.*

When $N = 3$, the formula becomes the following (we denote by $l(a)$ the length of the edge a and by $\angle(a)$ the dihedral angle of the edge a):

$$\mathrm{Vol}_3(\mathcal{K}_\varepsilon) = \mathrm{Vol}_3(\mathcal{K}) + A(\partial \mathcal{K})\varepsilon + \left(\sum_{a \in \partial K} \angle(a)l(a) \right)\varepsilon^2 + \frac{4}{3}\pi\varepsilon^3. \tag{16.13}$$

Sketch of proof of Theorem 47 We shall decompose $\mathcal{K}_\varepsilon$ into a union of subsets "lying above the k-faces" ($0 \le k \le N$) and evaluate the volume of each portion. Note that

$$\partial \mathcal{K} = \cup_{0 \le k \le N-1} \mathcal{F}^k,$$

---

[1] This angle has been defined in Chap. 6.

where $\mathcal{F}^k$ is the union of the k-faces of $\partial \mathcal{K}$. Let σ^k be any k-face of $\partial \mathcal{K}$. We define

$$B_k(\sigma^k, \varepsilon) = \cup_{\xi \in L_m, m \in \sigma^k}[m, m + \varepsilon \xi],$$

where L_m denotes the basis of the normal cone $C^{\perp}(\sigma^k)$ at any point m of σ^k (see Chap. 6). Note that

$$\mathrm{Vol}_N(B_k(\sigma^k, \varepsilon)) = \mathrm{Vol}_N(L(\sigma^k))\varepsilon^k.$$

On the other hand,

$$\mathcal{K}_\varepsilon = \mathcal{K} \cup_{0 \leq k \leq N-1} \cup_{\sigma^k \in \partial \mathcal{K}} B_k(\sigma^k, \varepsilon).$$

Since the intersection of two portions of the previous union is empty or has a null N-measure, we deduce that

$$\mathrm{Vol}_N(\mathcal{K}_\varepsilon) = \mathrm{Vol}_N(\mathcal{K}) + \sum_{0 \leq k \leq N-1} \mathrm{Vol}_N(B_k(\sigma^k, \varepsilon)),$$

from which we get (16.9). Clearly,

$$\Phi_0(\mathcal{K}) = \mathrm{Vol}_N(\mathcal{K}), \text{ and } \Phi_1(\mathcal{K}) = \mathrm{Vol}_{N-1}(\partial \mathcal{K}),$$

since for any $(N-1)$-face σ^{N-1} and for any $m \in \sigma^{N-1}$ L_m is reduced to a point.

Finally, the union of the basis $L(\sigma^0)$ over all 0-simplices is an $(N-1)$-sphere $\mathbb{S}^{N-1}$, from which we obtain

$$\Phi_N(\mathcal{K}) = \mathrm{Vol}_N(\mathbb{B}^N).$$

Details and further results can be found in [9, Tome 3, 12.2, 12.3]. □

16.4 Convex Bodies with Smooth Boundary

Now, we deal with the case where $\mathcal{K}$ is a convex body with smooth boundary $\partial \mathcal{K}$ (once more, we can reduce our study to the case of a body using (16.8)). We shall compute the coefficients of the *Steiner polynomial*. Using classical differential geometry, we prove that the coefficients $\Phi_k(\mathcal{K})$ are proportional to the (global) $(k-1)^{th}$-mean curvatures (also called *global Lipschitz–Killing curvatures*) $\mathbb{M}_{k-1}(\partial \mathcal{K})$ of $\partial \mathcal{K}$, defined by

$$\mathbb{M}_r(\partial \mathcal{K}) = \frac{1}{C_{N-1}^r} \int_{\partial \mathcal{K}} \Xi_k dv_{\partial \mathcal{K}}, \tag{16.14}$$

Ξ_k denoting the kth-elementary symmetric function of the principal curvatures $\lambda_1, ..., \lambda_{N-1}$ of $\partial \mathcal{K}$. The *Steiner formula* can be stated as follows.

Theorem 48. *Let $\mathcal{K}$ be a convex body in $\mathbb{E}^N$ whose boundary $\partial\mathcal{K}$ is a hypersurface of class C^2. Then*

$$\mathrm{Vol}_N(\mathcal{K}_\varepsilon) = \sum_{k=0}^{k=N} \Phi_k(\mathcal{K})\varepsilon^k, \tag{16.15}$$

with

$$\Phi_0(\mathcal{K}) = \mathrm{Vol}_N(\mathcal{K}), \tag{16.16}$$

and, for each $k \geq 1$,

$$\Phi_k(\mathcal{K}) = \frac{C_N^k}{N} \mathbb{M}_{k-1}(\partial\mathcal{K}). \tag{16.17}$$

Thus, the *Steiner formula* can be stated as follows:

$$\mathrm{Vol}_N(\mathcal{K}_\varepsilon) = \mathrm{Vol}_N(\mathcal{K}) + \sum_{k=1}^{N} \frac{C_N^k}{N}\mathbb{M}_{k-1}(\partial\mathcal{K})\varepsilon^k, \forall \varepsilon \geq 0. \tag{16.18}$$

In particular, if $\mathcal{K}$ is a convex body in $\mathbb{E}^3$ with smooth boundary, then

$$\mathrm{Vol}_3(\mathcal{K}_\varepsilon) = \mathrm{Vol}_3(\mathcal{K}) + A(\partial\mathcal{K})\varepsilon + (\int_{\partial\mathcal{K}} Hda)\varepsilon^2 + \frac{1}{3}(\int_{\partial\mathcal{K}} Gda)\varepsilon^3, \tag{16.19}$$

where H (resp., G) denotes the mean curvature (resp., Gauss curvature) of $\partial\mathcal{K}$.

Note that the Gauss–Bonnet theorem implies that the last term is equal to $\frac{4}{3}\pi\varepsilon^3$ since $\mathcal{K}$ is convex. Consequently, we can state that

$$\mathrm{Vol}_3(\mathcal{K}_\varepsilon) = \mathrm{Vol}_3(\mathcal{K}) + A(\partial\mathcal{K})\varepsilon + (\int_{\partial\mathcal{K}} Hda)\varepsilon^2 + \frac{4}{3}\pi\varepsilon^3. \tag{16.20}$$

Proof of Theorem 48 Let

$$x : \partial\mathcal{K} \to \mathbb{E}^N$$

be an isometric immersion of $\partial\mathcal{K}$ into $\mathbb{E}^N$. If ξ denotes a generic (outward) unit normal vector field, then

$$\phi(t) = x + t\xi$$

generates $\partial\mathcal{K}_t$ in $\mathbb{E}^N$.

Let $(e_i, 1 \leq i \leq N-1)$ be a local frame on $\partial\mathcal{K}$. Without loss of generality, we can assume that they are principal vectors of $\mathcal{K}$ in the direction ξ. For every $i, 1 \leq i \leq N-1$, we have

$$A_\xi e_i = \lambda_i^\xi e_i,$$

where λ_i^ξ denotes the principal curvatures of $\mathcal{K}$ (with respect to ξ). Then

$$d\phi_t(e_i) = e_i + A_{t\xi}e_i = e_i + t\lambda_i^\xi e_i.$$

Consequently, the volume form $dv_{\partial \mathcal{K}_t}$ of $\partial \mathcal{K}_t$ satisfies

$$dv_{\partial \mathcal{K}_t} = (\omega^1 + t\lambda_1^\xi \omega^1) \wedge \ldots \wedge (\omega^{N-1} + t\lambda_{N-1}^\xi \omega^{N-1}) =$$

$$\sum_k \Xi_k(\xi) t^k \omega^1 \wedge \ldots \wedge \omega^{N-1},$$

where $(\omega^i, 1 \leq i \leq N-1)$ is the dual frame of $(e_i, 1 \leq i \leq N-1)$. Then

$$\mathrm{Vol}_N(\mathcal{K}_\varepsilon) = \mathrm{Vol}_N(\mathcal{K}) + \sum_k \int_{t=0}^{t=\varepsilon} \Xi_k(\xi) dv_{\partial \mathcal{K}_t} t^k dt,$$

i.e.,

$$\mathrm{Vol}_N(\mathcal{K}_\varepsilon) = \mathrm{Vol}_N(\mathcal{K}) + \sum_{k=1}^{N} \frac{C_N^k}{N} \mathbb{M}_{k-1}(\partial \mathcal{K}),$$

which implies (16.18).

Details and further results can be found in [73, Chap. 13, Sect. 5]. □

16.5 Evaluation of the Quermassintegrale by Means of Transversal Integrals

We have seen in the previous sections that the *Quermassintegrale* can be evaluated in some particular cases in terms of local invariants: angles, lengths for polyhedra, and principal curvatures when the boundary is smooth. We give here a *global* interpretation of these quantities by means of *transversal integrals* introduced in Definition 24, using the Cauchy and Kubota formulas (see Theorems 27 and 29).

Theorem 49. *Let $\mathcal{K}$ be a convex body of $\mathbb{E}^N$. Then, for every k,*

$$\Phi_k(\mathcal{K}) = C_N^k \mathbb{W}_k(\mathcal{K}), \tag{16.21}$$

where $\mathbb{W}_k$ denotes the kth-transversal integral of $\mathcal{K}$.

Consequently, the *Steiner formula* can be stated as follows:

$$\forall \varepsilon \geq 0, \mathrm{Vol}_N(\mathcal{K}_\varepsilon) = \sum_{k=0}^{N} C_N^k \mathbb{W}_k(\mathcal{K}) \varepsilon^k. \tag{16.22}$$

Proof of Theorem 49 We prove this theorem by induction on the dimension N of the ambient space:

- For $N = 1$, the result is trivial.
- We assume that the result is true up to $N-1$. In particular, for each hyperplane P^{N-1}, we have

$$\forall \rho \geq 0, \operatorname{Vol}_{N-1}(\mathcal{K}_{\rho_{pN-1}}) = \sum_{k=0}^{N-1} C_N^k \mathbb{W}_k(\mathcal{K}_{P^{N-1}}) \rho^k. \qquad (16.23)$$

On the other hand,

$$\operatorname{Vol}_N(\mathcal{K}_\varepsilon) = \operatorname{Vol}_N(\mathcal{K}) + \int_0^\varepsilon \operatorname{Vol}_{N-1}(\partial \mathcal{K}_\rho) d\rho. \qquad (16.24)$$

Using the Cauchy formula (Theorem 27), we get

$$\operatorname{Vol}_{N-1}(\partial \mathcal{K}_\rho) = \int_{G(N,N-1)} \operatorname{Vol}_{N-1}(\mathcal{K}_{\rho_{pN-1}}) dP^{N-1}.$$

Using induction, we have

$$\int_{G(N,N-1)} \operatorname{Vol}_{N-1}(\mathcal{K}_{\rho_{pN-1}}) P^{N-1} = \sum_0^{N-1} C_N^k \int_{G(N,N-1)} \mathbb{W}_k(\mathcal{K}_{P^{N-1}}) \rho^k P^{N-1}.$$

By the *Kubota formula* (Theorem 29), we get

$$\int_{G(N,N-1)} \operatorname{Vol}_{N-1}(\mathcal{K}_{\rho_{pN-1}}) P^{N-1} = \sum_0^{N-1} C_N^k \mathbb{W}_{k+1}(\mathcal{K}_{P^{N-1}}) \rho^k.$$

We deduce that

$$\forall \varepsilon \geq 0, \operatorname{Vol}_N(\mathcal{K}_\varepsilon) = \sum_{k=0}^{N} C_N^k \mathbb{W}_k(\mathcal{K}) \varepsilon^k. \quad \square \qquad (16.25)$$

16.6 Continuity of the Φ_k

It is interesting to note that the coefficients Φ_k satisfy continuity properties. We have already seen that Φ_0 (i.e., the N-volume) and Φ_1 (i.e., the $(N-1)$-volume of the boundary) are continuous on the set $\mathcal{C}_b$ of convex bodies of $\mathbb{E}^N$ endowed with the Hausdorff topology (see Theorem 26). Since the other Φ_k are proportional to the transversal integrals $\mathbb{W}_k$, which are *continuous* valuations (see Proposition 6), they are themselves continuous.

As an application of the *Steiner* formula (Theorem 46), we shall be more precise as follows.

Theorem 50. *Let $\mathcal{K}$ and $\mathcal{K}'$ be convex bodies of $\mathbb{E}^N$ whose Hausdorff distance is smaller than ρ. Then for each $k, 0 \leq k \leq N$, there exists a constant $C_k(N,\mathcal{K})$ (depending only on the Quermassintegrale of $\mathcal{K}$ and the dimension N of the ambient space) such that*

$$|\Phi_k(\mathcal{K}) - \Phi_k(\mathcal{K}')| \leq C_k(N,\mathcal{K})\rho. \qquad (16.26)$$

Sketch of proof of Theorem 50 The proof is by induction on k. For $k=0$, the result is a consequence of Theorem 46: since $\mathcal{K}' \in \mathcal{K}_\rho$,

$$|\mathrm{Vol}_N\mathcal{K}' - \mathrm{Vol}_N\mathcal{K}| \leq |\mathrm{Vol}_N\mathcal{K}_\rho - \mathrm{Vol}_N\mathcal{K}|.$$

Using the Steiner formula, one has

$$|\mathrm{Vol}_N\mathcal{K}_\rho - \mathrm{Vol}_N\mathcal{K}| \leq C_0(N,\mathcal{K})\rho,$$

where $C_0(N,\mathcal{K})$ is a constant depending on the dimension of the ambient space and the Quermassintegrale of $\mathcal{K}$. Consequently, the first part of Theorem 50 is proved for $k=0$. One completes the proof by induction, using the Kubota formula (Theorem 29), since the coefficients Φ_k are proportional to the coefficients $\mathbb{W}_k$. □

Remarks

- It is important to note that Theorem 50 gives *quantitative informations*: the knowledge of the geometry of $\mathcal{K}$ and the Hausdorff distance between $\mathcal{K}$ and $\mathcal{K}'$ give a bound on the difference of the transversal integrals.
- It must also be noted that *the assumption of convexity* is crucial to obtain continuity results for every Φ_k. For instance, we have seen in Chap. 3 the well-known example of the *Lantern of Schwarz* in $\mathbb{E}^3$, which shows that the area is not continuous with respect to the Hausdorff topology.
- As an obvious consequence of Theorem 50, if $(\mathcal{K}_n)_{n\geq 0}$ is a sequence of convex bodies of $\mathbb{E}^N$ whose Hausdorff limit is the (compact) convex subset $\mathcal{K}$, then

$$\forall k \geq 0, \lim_{n\to\infty} \Phi_k(\mathcal{K}_n) = \Phi_k(\mathcal{K}). \tag{16.27}$$

Note that the dimension of $\mathcal{K}$ may be different to the dimension of $\mathcal{K}_n$. For instance, a sequence of "full cylinders" with the same axis and radius $\frac{1}{n}$ may tend to a straight line. In this case, the limit is not a body and one can use directly (16.5) to compute its *Quermassintegrale*.

- In $\mathbb{E}^3$, the only interesting information given by Theorem 50 is the behavior of the mean curvature of closed surfaces: we deduce from (16.13) and (16.20) the following result.

Corollary 12. *Let P_n be a sequence of convex polyhedra in $\mathbb{E}^3$ whose Hausdorff limit is a (closed convex) surface S. Then, the mean curvature H of S satisfies*

$$\int_S Hda = \lim_{n\to\infty} \sum_{a_n} \angle(a_n)l(a_n) \tag{16.28}$$

($l(a_n)$ denoting the length of the edge a_n, $\angle(a_n)$ the dihedral angle of the edge a, and H the mean curvature of S).

Warning! Noncontinuity of Pointwise Curvatures. At this step, it must be noticed that, although we have a *global continuity result* on the collection of convex subsets of $\mathbb{E}^N$, there is nonpointwise continuity. For instance, consider the hypersphere $\mathbb{S}^{N-1} \subset \mathbb{E}^N$, a point m on it, and the normal line ξ_m of $\mathbb{S}^{N-1}$ at m. Let $\mathcal{K}_n$ be a sequence of (convex) polyhedra which tend to unit ball B^N for the Hausdorff topology, such that the sequence m_n of points obtained by intersecting the normal line ξ_m with $\partial \mathcal{K}_n$ are smooth points of $\partial \mathcal{K}_n$. The curvatures of $\partial \mathcal{K}_n$ at m_n are obviously null since the $\partial \mathcal{K}_n$ are totally geodesic at the interior points of their $(N-1)$-faces, although the curvatures at every point of $\mathbb{S}^{N-1}$ are 1.

16.7 An Additivity Formula

It is also interesting to note that each *Quermassintegrale* satisfies an additivity formula. Indeed, if $\mathcal{K}_1$ and $\mathcal{K}_2$ are two convex bodies of $\mathbb{E}^N$, then their N-volumes satisfy

$$\mathrm{Vol}_N(\mathcal{K}_1 \cup \mathcal{K}_2) = \mathrm{Vol}_N(\mathcal{K}_1) + \mathrm{Vol}_N(\mathcal{K}_2) - \mathrm{Vol}_N(\mathcal{K}_1 \cap \mathcal{K}_2). \qquad (16.29)$$

(This formula is obviously true even if $\mathcal{K}_1$ and $\mathcal{K}_2$ are not convex since the N-volume $\mathrm{Vol}_N(\mathcal{K})$ is nothing but the Lebesgue measure of the subset $\mathcal{K}$.)

Moreover, we have seen in Proposition 6 that every transversal integral $\mathbb{W}_k$ is a valuation and then is additive. Consequently,

$$\mathbb{W}_k(\mathcal{K}_1 \cup \mathcal{K}_2) = \mathbb{W}_k(\mathcal{K}_1) + \mathbb{W}_k(\mathcal{K}_2) - \mathbb{W}_k(\mathcal{K}_1 \cap \mathcal{K}_2) \qquad (16.30)$$

and

$$\Phi_k(\mathcal{K}_1 \cup \mathcal{K}_2) = \Phi_k(\mathcal{K}_1) + \Phi_k(\mathcal{K}_2) - \Phi_k(\mathcal{K}_1 \cap \mathcal{K}_2). \qquad (16.31)$$

Chapter 17
Tubes Formula

In Chap. 16, we have seen that the volume of the parallel body of a convex body with smooth boundary is a polynomial whose coefficients depend on the second fundamental form of the boundary. This formula has been generalized by Weyl [82] for the volume of tubes around any smooth submanifold in $\mathbb{E}^N$, with or without boundary.

17.1 The Lipschitz–Killing Curvatures

In Chap. 11, we have defined the notion of kth-mean curvature of a submanifold M^n of $\mathbb{E}^N$ of any codimension. Using the Gauss equation (11.17) and integration over the fibers of the unit normal bundle (see Theorem 32), Cheeger et al. [25] proved that these quantities are related to the so-called *Lipschitz–Killing curvatures* of M^n, which are intrinsic invariants of M^n, i.e., depending only on the curvature tensor of M^n and the second fundamental form of its eventual boundary.[1]

Let us begin with the precise definition of the Lipschitz–Killing curvatures.

Definition 40. Let M^n be a Riemannian manifold (with or without boundary), isometrically embedded in $\mathbb{E}^N$:

1. At each point of $M^n \backslash \partial M^n$, the jth Lipschitz–Killing curvature R_j is the n-differential form locally defined as follows:

   - If j is odd, R_j is null.
   - If j is even, R_j is given by

$$R_j = (-1)^{j/2}((n-j)!2^j \pi^{j/2}(j/2)!)^{-1} \sum_{\sigma \in \mathcal{S}^n} (-1)^{\sigma_{i_1 \dots i_n}} \Omega_{i_2}^{i_1} \wedge \dots \wedge \Omega_{i_j}^{i_{(j-1)}}$$

$$\wedge \, \omega^{i_{(j+1)}} \wedge \dots \wedge \omega^{i_n}. \tag{17.1}$$

---

[1] The boundary is considered as a hypersurface of M^n.

2. Similarly, the jth Lipschitz–Killing curvature of ∂M^n in M^n is the $(n-1)$-differential form locally defined by

$$H_j = \sum_k Q_{kj}, \tag{17.2}$$

where

$$Q_{kj} = c_{kj} \sum_{\sigma \in \mathscr{S}^{n-1}} (-1)^{\sigma_{i_1 \cdots i_{(n-1)}}} \Omega^{i_1}_{i_2} \wedge \ldots \wedge \Omega^{i_{(2k-1)}}_{i_{2k}}$$
$$\wedge \, \omega^{i_{(2k+1)}}_n \wedge \ldots \wedge \omega^{i_{(j-1)}}_n \wedge \omega^{i_j} \wedge \ldots \wedge \omega^{i_{n-1}}, \tag{17.3}$$

with

$$c_{kj} = \begin{cases} (-1)^{k+1}((n-j)!2^j\pi^{(j-1)/2}k!(\frac{j-1}{2}-k)!)^{-1}, & \text{if } j \text{ is odd,} \\ (-1)^k((n-j)!2^{k+j/2}\pi^{j/2}k!(j-2k-1)!!)^{-1}, & \text{if } j \text{ is even,} \end{cases}$$

where $(j-2k-1)!! = 1.3.5....(j-2k-1)$.

(The 1-forms $(\omega^1,...,\omega^n)$ are the dual forms of a local orthonormal frame $(e_1,...,e_n)$ such that e_n is normal to ∂M^n; the 1-forms ω^j_i are the corresponding connection forms; and the 2-forms Ω^i_j are the curvature forms of M^n. The sum is over all the permutations σ of n elements.)

In particular:

- R_0 is the volume form of M^n.
- R_n is null if n is odd, and proportional to $\sum_\sigma \Omega^{i_1}_{i_2} \wedge \ldots \wedge \Omega^{i_{n-1}}_{i_n}$ if n is even.

The form R_n is involved in the Gauss–Bonnet theorem as we shall see later.

The link between the notion of kth-mean curvature and k^{th} Lipschitz–Killing curvature is given by the following proposition (in which we identify $ST^\perp_m M^n$ with $\mathbb{S}^{N-n-1}$ for every $m \in M^n \setminus \partial M^n$ and $ST^\perp_m \partial M^n$ with $\mathbb{S}^{N-n}$ for every m in ∂M^n).

Proposition 13. *Let M^n be a compact oriented n-dimensional Riemannian manifold (with or without boundary), isometrically embedded in $\mathbb{E}^N$:*

1. Let m be a (fixed) point of $(M^n \backslash \partial M^n)$. Then, $\Xi_k(\xi)_m$ is a function defined on $ST^\perp_m M^n$ such that:

- *If k is odd,*

$$\int_{\mathbb{S}^{N-n-1}} \Xi_k(\xi)d\xi_m = 0. \tag{17.4}$$

- *If k is even,*

$$(\int_{\mathbb{S}^{N-n-1}} \Xi_k(\xi)d\xi_m)dv_M = R_k(m), \tag{17.5}$$

this equality being considered as an equality between two n-forms at each point m of $(M^n \setminus \partial M^n)$.

2. Let m be a (fixed) point of ∂M^n. Then, $\Xi_k(\xi)_m$ is a function defined on $ST_m^\perp \partial M^n = \mathbb{S}^{N-n}$ such that

$$(\int_{\mathbb{S}^{N-n}} \Xi_k(\xi) d\xi_m) dv_{\partial M^n} = H_k(m), \qquad (17.6)$$

this equality being considered as an equality between two $(n-1)$-forms at each point of ∂M^n.

Sketch of proof of Proposition 13 We indicate the proof for a point $m \in (M^n \setminus \partial M^n)$:

1. If k is odd (and for a fixed $m \in (M^n \setminus \partial M^n)$),

$$\int_{\mathbb{S}^{N-n-1}} \Xi_k(\xi) d\xi_m = 0, \qquad (17.7)$$

since in this case $\Xi_k(-\xi)_m = -\Xi_k(\xi)_m$.
2. If k is even (and for a fixed $m \in (M^n \setminus \partial M^n)$),

$$(\int_{\mathbb{S}^{N-n-1}} \Xi_k(\xi) d\xi_m) dv_{\partial M^n} \qquad (17.8)$$

can be expressed in terms of the curvature forms of M^n. Indeed, consider (17.1):

$$R_j = (-1)^{j/2}((n-j)!2^j \pi^{j/2}(j/2)!)^{-1} \sum_{\sigma \in \mathcal{S}^n} (-1)^{\sigma_{i_1 \dots i_n}} \Omega_{i_2}^{i_1} \wedge \dots \wedge \Omega_{i_j}^{i_{(j-1)}}$$

$$\wedge \omega^{i_{(j+1)}} \wedge \dots \wedge \omega^{i_n}.$$

Each term Ω_j^i can be expressed with help of the Gauss equation:

$$\Omega_j^i = \sum_\alpha \omega_\alpha^i \wedge \omega_j^\alpha, \qquad (17.9)$$

where the sum is taken over any normal frame $\{\xi_1, \dots, \xi_\alpha, \dots, \xi_{(N-n)}\}$. Moreover, each term ω_i^α can be expressed with help of the tensor h_{ij}^α:

$$\omega_i^\alpha = \sum_j h_{ij}^\alpha \omega^j.$$

Consequently, (17.1) becomes

$$R_j = (-1)^{j/2}((n-j)!2^j \pi^{j/2}(j/2)!)^{-1} \sum_{\sigma \in \mathcal{S}^j}(-1)^{\sigma_{i_1 \dots i_j}} (\sum_{\alpha,l} h_{i_1 l}^\alpha h_{i_2 l}^\alpha)$$
$$\dots (\sum_{\alpha,l} h_{i_{(j-1)}l}^\alpha h_{i_j l}^\alpha) \omega^{i_1} \wedge \omega^{i_2} \wedge \dots \wedge \omega^{i_j} \wedge \omega^{i_{(j+1)}} \wedge \dots \wedge \omega^{i_n}. \qquad (17.10)$$

On the other hand, if ξ is any normal vector, $\Xi_k(\xi)$ is the sum of all determinants of k-minors of the matrix h_{ij}^ξ. One can decompose ξ into the frame ξ_α. Integrating on the sphere $\mathbb{S}^{N-n-1}$, it is easy to conclude that the only non-null terms in the sum are exactly the R_k.

3. Similar proofs apply for a point on ∂M^n. □

17.2 The Tubes Formula of Weyl (1939)

Let M^n be a Riemannian manifold (with or without boundary), isometrically embedded in $\mathbb{E}^N$. We shall evaluate the volume of the tube M^n_ε of M^n of radius ε.

17.2.1 The Volume of a Tube

Using the previous Lipschitz–Killing forms, we introduce now the *global Lipschitz–Killing curvatures* $\mathbb{M}_k$.

Definition 41. The *global Lipschitz–Killing curvatures* $\mathbb{M}_k$ of a (compact) submanifold M^n of $\mathbb{E}^N$ are defined by

$$\mathbb{M}_k(M^n) = \begin{cases} \frac{s_{k-1}}{C_N^k}\left(\int_{M^n} R_{(k+n-N)} + \int_{\partial M^n} H_{(k+n-N)}\right), & \text{if } k+n-N \ge 0, \\ \mathbb{M}_k(M^n) = 0 & \text{otherwise.} \end{cases} \tag{17.11}$$

Of course, these quantities coincide with those defined in the case of a convex hypersurface (see (16.14)).

We can now state the theorem of Weyl, which is an easy consequence of Proposition 13.

Theorem 51. *Let M^n be a compact oriented n-dimensional Riemannian manifold (with or without boundary), isometrically embedded in $\mathbb{E}^N$. Then, the volume $\mathrm{Vol}_N(M^n_\varepsilon)$ of M^n_ε is a polynomial in ε:*

$$\mathrm{Vol}_N(M^n_\varepsilon) = \sum_{k=0}^{N} C_N^k \mathbb{M}_k(M^n)\varepsilon^k. \tag{17.12}$$

Using the notation of Chap. 16, Theorem 51 implies that one can write

$$\mathrm{Vol}_N(M^n_\varepsilon) = \sum_{k=0}^{N} \Phi_k(M^n)\varepsilon^k, \text{with } \Phi_k(M^n) = C_N^k \mathbb{M}_k(M^n). \tag{17.13}$$

Remarks

1. In particular, we see that the first nonzero Φ_k is Φ_{N-n}.
2. Suppose that M^n is a (compact oriented) *n*-dimensional Riemannian manifold without boundary, isometrically embedded in $\mathbb{E}^N$. Then, we deduce from

the definition of the scalar curvature (see Chap. 10) that $\Phi_{N-n+2}(M^n)$ equals $\int_{M^n} r \, dv_{M^n}$, up to a constant depending on the dimensions, where r denotes the scalar curvature of M^n.

3. With the same assumptions, and assuming that n is even, we have

$$\Phi_N(M^n) = C \int_{M^n} \sum_{\sigma \in \mathcal{S}^n} (-1)^{\sigma_{i_1 \cdots i_n}} \Omega_{i_2}^{i_1} \wedge \ldots \wedge \Omega_{i_n}^{i_{(n-1)}}, \qquad (17.14)$$

with $C = \dfrac{(-1)^{n/2} C_N^n s_{N-1}}{(N-n)! \, 2^n \pi^{n/2} (n/2)!}$.

4. With the same assumptions,

$$\Phi_{N-n}(M^n) = s_{N-1} \operatorname{Vol}_n(M^n). \qquad (17.15)$$

5. In particular, if M^n has no boundary, then the Φ_{N-n+k} are null when k is odd.

6. In this section, we have supposed for simplicity that M^n is oriented. However, $ST^\perp M^n$ is always oriented even if M^n is not. This can be seen as follows. We can identify ∂M_ε^n with $ST^\perp M^n$, via the exponential map (modified by a factor ε):

$$\phi : (m + \varepsilon \xi_m) \to (m, \xi_m).$$

Since ∂M_ε^n is a closed hypersurface of $\mathbb{E}^N$, it has a canonical orientation given by the outward normal. The manifold $ST^\perp M^n$ can be endowed with the orientation given by this identification.

7. The exact values of the constant coefficients have been computed in [24, 25]. We leave it to the reader to check them.

Sketch of proof of Theorem 51 We give here a short idea of the proof (when the submanifold has no boundary). Let

$$x : M^n \to \mathbb{E}^N$$

be the isometric immersion of M^n into $\mathbb{E}^N$. If ξ denotes a generic unit normal vector, then $\phi(t) = x + t\xi$ generates the sphere normal bundle $S_t T^\perp M^n$ of radius t, considered as a hypersurface of $\mathbb{E}^N$ (Fig. 17.1).

Let $(e_i)_{1 \leq i \leq n}$ be a local frame of principal vectors in the direction ξ. For every $i(1 \leq i \leq n)$, we have

$$A_\xi e_i = \lambda_i^\xi e_i.$$

Then

$$d\phi_t(\varepsilon_i) = e_i + A_{t\xi} e_i = e_i + t\lambda_i^\xi e_i.$$

Consequently, the volume form $dv_{S_t T^\perp M^n}$ of the sphere normal bundle $S_t T^\perp M^n$ of radius t at the point ξ satisfies

$$\begin{aligned} dv_{S_t T^\perp M^n} &= (\omega_1 + t\lambda_1^\xi \omega^1) \wedge \ldots \wedge (\omega^n + t\lambda_n^\xi \omega^n) \wedge t^{N-n-1} dv_{S^{N-n-1}} \\ &= \sum_k \Xi_k(\xi) t^k \omega^1 \wedge \ldots \wedge \omega^n \wedge t^{N-n-1} dv_{S^{N-n-1}}. \end{aligned}$$

Fig. 17.1 The map $m \to m + t\xi$

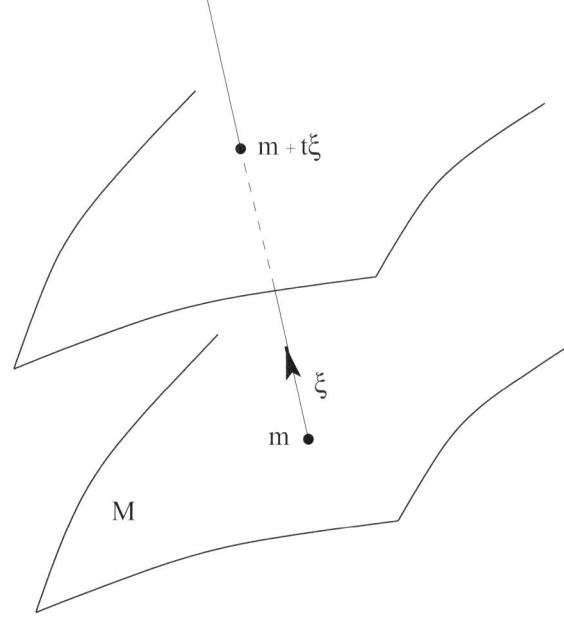

Then, using the area formula (Theorem 19), we find that

$$\mathrm{Vol}_N(M_\varepsilon^n) = \sum_k \int_{t=0}^{\varepsilon} \int_{ST^{\perp}M^n} \Xi_k(\xi) dv_{S_t T^{\perp}M^n} t^k dt$$

$$= \int_{t=0}^{t=\varepsilon} \int_{S^{N-n-1}} \int_{M^n} \sum_k \Xi_k(\xi) t^{N-n-1+k} \omega^1 \wedge \ldots \wedge \omega^n \wedge dt \wedge dv_{S^{N-n-1}}$$

$$= \frac{(N-k)!}{n!} k! s_{k-1} (\int_{M^n} \Xi_k(\xi) dv_M) \varepsilon^{N-n+k},$$

$$(17.16)$$

where (ω^i) denotes the dual frame of e_i.

The left-hand side of (17.16) is a polynomial in ε, whose odd terms are null by Proposition 13 since we assume here that M^n has no boundary. The even terms can be written in terms of R_j by using the Gauss equation (17.9), from which we deduce Theorem 51. The proof is similar for submanifolds with boundary. $\square$

17.2.2 Intrinsic Character of the $\mathbb{M}_k$

Weyl made the following fundamental remark, which is an obvious consequence of our previous computation: at each point of M^n, the Lipschitz–Killing curvature forms *depend only on the intrinsic geometry of M^n* and are independent of the isometric embedding. They can be computed with only the knowledge of the curvature tensor of M^n and the second fundamental form of the boundary ∂M^n, considered as a hypersurface of M^n. This implies that two isometric immersions of the same manifold M^n in $\mathbb{E}^N$ have the same Lipschitz–Killing curvatures. In particular:

- The quantities $\mathbb{M}_k$ depend only on the intrinsic geometry of M^n and are consequently independent of the embedding.
- Moreover, as for the coefficients $\Lambda_{k,N}$ defined in the convex case (see (16.6), Chap. 16), they are independent of the codimension. Once again, the Lipschitz–Killing curvature forms and the global Lipschitz–Killing curvatures are intrinsic invariants of M^n.
- When M^n has no boundary, the only non-null terms $\mathbb{M}_k$ are the even ones. These coefficients are the same as those which appear in the *Gauss–Bonnet formula*, generalized by Allendoerfer and Weil [3] and Chern [28] among others.

17.3 The Euler Characteristic

Recall that the Gauss–Bonnet theorem relates the Euler characteristic of a manifold with its curvature tensor (see Chap. 10). If M^n is even dimensional and has no boundary,

$$\chi(M^n) = \frac{(-1)^{n/2}}{2^n \pi^{n/2} (n/2)!} \int_{M^n} \sum_{\sigma \in \mathcal{S}^n} (-1)^{\sigma_{i_1 \dots i_n}} \Omega^{i_1}_{i_2} \wedge \dots \wedge \Omega^{i_{(n-1)}}_{i_n}. \qquad (17.17)$$

With our notation, (17.14) implies that

$$\Phi_N(M^n) = \mathbb{M}_N(M^n) = s_{N-1} \int_{M^n} R_n = b_N \chi(M^n). \qquad (17.18)$$

If M^n has a boundary, we add the boundary term in H_n:

$$\Phi_N(M^n) = \mathbb{M}_N(M^n) = s_{N-1} \int_{M^n} R_n + \int_{\partial M^n} H_n = b_N \chi(M^n). \qquad (17.19)$$

Note the important fact that the previous equations show that the term Φ_N has a topological meaning, since the Euler characteristic is a topological invariant. This result can be seen as a generalization of the convex case, since the Euler characteristic of a convex subset is always equal to 1.

17.4 Partial Continuity of the Φ_k

It is clear that the continuity of the Φ_k mentioned for convex bodies is not satisfied for smooth submanifolds: for instance, using a construction analogous to the *Lantern of Schwartz* (Sect. 3.1.3), it is easy to construct an example of sequence of smooth surfaces, which approaches a finite cylinder and whose areas tend to infinity. However, adding suitable assumptions on the sequence of submanifolds (in particular by bounding the principal curvatures), one can obtain convergence results.

Theorem 52. *Let $\varepsilon > 0$. Let $\mathcal{S}^\varepsilon$ be the class of n-dimensional compact submanifolds M^n of $\mathbb{E}^N$ whose second fundamental form is bounded by ε. Then for every k, the map*

$$M^n \to \Phi_k(M^n)$$

is continuous on $\mathcal{S}^\varepsilon$, endowed with the Hausdorff topology.

A complete proof is left to the reader. It can be found in [42].

17.5 Transversal Integrals

In Chap. 16, we have seen that one has the possibility of expressing the coefficients $\Phi_k(\mathcal{K})$ in terms of *transversal integrals* when $\mathcal{K}$ is convex. This can be done also for the tubes formula of any smooth submanifold, in terms of the Euler characteristic of the intersection of the submanifold with affine subspaces (see Sect. 10.7.3). The following formula is a direct generalization of the convex case, since the Euler characteristic of a convex body is 1.

Theorem 53. *Let M^n be a (compact oriented) submanifold embedded in $\mathbb{E}^N$. Then, for every $k, 0 \le k \le N$,*

$$\Phi_k(M^n) = C_N^k \frac{b_N}{b_{N-k}\gamma_{N,k}} \int_{AG(N,k)} \chi(M^n \cap P^k)dP^k, \qquad (17.20)$$

where $AG(N,k)$ denotes the Grassmann manifold of affine k-subspaces of $\mathbb{E}^N$.

Proof of Theorem 53 This formula is a special case of the kinematic formula of Chern [30]. The proof we present here is that of Chern, adapted to the linear case. We refer to [30] for details. Consider the subset of affine k-subspaces P^k whose intersection with M^n is nonempty. Generically, such an intersection is a submanifold of dimension $n + k - N$ (one can restrict attention to the open set of affine k-subspaces having such an intersection, since the union of the other ones has a null measure in the affine Grassmann manifold). Over $M^n \cap P^k$, consider a local orthonormal frame $(e_1, ..., e_N)$ of $\mathbb{E}^N$ such that $(e_1, ..., e_k)$ is tangent to P^k and $(e_1, ..., e_{n+k-N})$ is tangent to $M^n \cap P^k$. Then, $(e_{n+k-N+1}, ..., e_N)$ is normal to $M^n \cap P^k$ and we shall introduce the "angle" between this frame and the tangent plane to M^n. Now, let $(\varepsilon_{k+1}, ..., \varepsilon_N)$ be orthonormal vector fields such that

$$(e_1, ..., e_{n+k-N}, \varepsilon_{k+1}, ..., \varepsilon_N)$$

is a local orthonormal frame of M^n. If ω_j denotes the dual frame of e_j, one can write

$$\omega_{k+\alpha} = \sum_{l=k+1}^{N} <\varepsilon_l, e_{k+\alpha}> \beta_l, 1 \leq \alpha \leq N-k,$$

with suitable 1-forms β_k. At a point $m \in M^n \cap P^k$, note that $\beta_{k+1} \wedge \ldots \wedge \beta_N$ is the volume form of the affine subspace orthogonal to P^k in M^n.

On the other hand, the volume form $dv_{\mathbb{E}^{N-k}}$ of the affine subspace orthogonal to P^k at m in $\mathbb{E}^N$ satisfies

$$dv_{\mathbb{E}^{N-k}} = \omega_{k+1} \wedge \ldots \wedge \omega_N = \Delta \beta_{k+1} \wedge \ldots \wedge \beta_N,$$

where $\Delta = \det <\varepsilon_l, e_{k+\alpha}>$. A crucial point is to note that Δ depends only on the mutual positions of P^k and $TM^n \cap P^k$. Moreover, at any point $m \in M^n \cap P^k$,

$$\beta_{k+1} \wedge \ldots \wedge \beta_N \wedge dv_{M^n \cap P^k} = dv_{M^n}.$$

Let Y be the manifold

$$\bigcup_{P^k \in AG(N,k)} (M^n \cap P^k, P^k).$$

The manifold Y can obviously be identified with $M^n \times G(N,k)$ and one can compare their volume form. Writing $dv_{AG(N,k)} = dv_{\mathbb{E}^{N-k}} \times dv_{G(N,k)}$, we deduce that

$$dv_{M^n \cap P^k} \wedge dv_{AG(N,k)} = \Delta dv_{M^n} \wedge dv_{G(N,k)}. \tag{17.21}$$

Multiplying (17.21) by $\overline{R}_{n+k-N}$, where $\overline{R}_{n+k-N}$ is the curvature form associated to the Riemannian manifold $M^n \cap P^k$, we get

$$\overline{R}_{n+k-N} dv_{M^n \cap P^k} \wedge dv_{AG(N,k)} = \overline{R}_{n+k-N} \Delta dv_{M^n} \wedge dv_{G(N,k)}. \tag{17.22}$$

Integrating the left-hand side of (17.22) over $AG(N,k)$ and using the Gauss–Bonnet theorem, we obtain

$$\int_{AG(N,k)} \int_{M^n \cap P^k} \overline{R}_{n+k-N} dv_{M^n \cap P^k} \wedge dv_{AG(N,k)} = c \int_{AG(N,k)} \chi(M^n \cap P^k) dv_{AG(N,k)}, \tag{17.23}$$

where c is a positive constant. To evaluate the right-hand side, we use Proposition 13: $\overline{R}_{n+k-N}$ can be written in terms of the second fundamental form $\overline{h}$ of $M^n \cap P^k$, considered as a submanifold of $\mathbb{E}^N$. Thus, at each point $m \in M^n \cap P^k$,

$$[\int_{\mathbb{S}^{N-(n+k-N)-1}} \Xi_{n+k-N}^{M^n \cap P^k}(\xi) d\xi] dv_{M^n \cap P^k} = \overline{R}_{n+k-N}. \tag{17.24}$$

Since P^k is totally geodesic in $\mathbb{E}^N$, $\overline{h}$ is related to the second fundamental form h of M^n by the equality

$$<\overline{h}(u,v), w> = \cos(\theta) <h(u,v), \xi>$$

for all vectors u, v tangent to $M^n \cap P^k$ and for all unit vectors ξ normal to M^n, w being the unit vector tangent to M^n and normal to $M^n \cap P^k$ defined by $(\cos\theta)w = \mathrm{pr}_{T M^n}\xi$. The other normal components of $\bar{h}$ are null.

Integrating the right-hand side of (17.22) over the product $M^n \times G(N,k)$, we remark that (17.22) implies that, at each point $m \in M^n$,

$$\int_{\xi\in\mathbb{S}^{2N-n-k-1}} \Xi^{M^n\cap P^k}_{n+k-N}(\xi)dv_{M^n\cap P^k} = c\int_{\xi\in\mathbb{S}^{N-n}} \Xi^{M^n}_{n+k-N}(\xi)dv_{M^n\cap P^k},$$

where $\mathbb{S}^{2N-n-k-1}$ denotes the unit sphere of vectors orthogonal to $T_m(M^n\cap P^k)$, $\mathbb{S}^{N-n}$ denotes the unit sphere of vectors orthogonal to $T_m(M^n\cap P^k)$ and tangent to P^k, and c is a constant depending only on the dimensions. Taking into account that Δ is independent of the point m in M^n, one finds

$$\int_{M^n}\int_{G(N,k)} \Delta\bar{R}_{n+k-N} = c\Phi_k(M^n),$$

where c is a suitable constant. This constant c can be evaluated by testing on spheres. Theorem 53 follows. $\square$

Remark Theorem 53 is a special case of the *kinematic formula* of Chern [30]. We state this formula in the following theorem, in which $G = SO(N)\alpha\mathbb{E}^N$ denotes the group of orientation-preserving Euclidean motions of $\mathbb{E}^N$.

Theorem 54. *Let M^n and N^k be two closed submanifolds of $\mathbb{E}^N$. Then,*

$$\int_{g\in G} \Phi_j(M^n\cap g(N^k))dg = \sum_{i \text{ even}} c_i\Phi_i(M^n)\Phi_{j-i}(N^k), \qquad (17.25)$$

where c_i are the constant, depending only on the dimensions.

17.6 On the Differentiability of the Immersions

In Sect. 17.5, we did not specify the differentiability of the involved submanifolds. Since the formulas need the curvature tensor, the natural framework is the space of immersions of class C^2. However, if the immersions are only C^1, but with Lipschitz normal frame, the same kind of formula can be proved, with slight modifications since the involved quantities have sense almost everywhere, Lipschitz maps being differentiable almost everywhere. The main results (obtained by integration) remain valid and the tubes formula can be written as follows:

$$\mathrm{Vol}_N(M^n_\varepsilon) = \int_{M^n}\left(\int_{\mathbb{S}^{N-n-1}}\left(\sum_k \Xi_k(M^n)dv_{\mathbb{S}^{N-n-1}}\right)\wedge\omega^1\wedge...\wedge\omega^n\right)\varepsilon^k, \qquad (17.26)$$

where Ξ_k is the coefficient of the term of degree k of the polynomial in the variable t. In particular, we still have

$$
\begin{aligned}
\Phi_k(M^n) &= 0, \forall k < N - n, \\
\Phi_{N-n}(M^n) &= s_{N-1} \operatorname{Vol}_n(M^n), \\
\Phi_N(M^n) &= s_{N-1} \chi(M^n).
\end{aligned}
\tag{17.27}
$$

Chapter 18
Subsets of Positive Reach

In previous chapters, we have seen that it is possible to define *curvatures* which describe the global shape of two classes of subsets of $\mathbb{E}^N$, namely the convex bodies and the smooth submanifolds. A good challenge is to find larger classes of subsets on which a more general theory holds. In 1958, Federer [43] made a major advance in two directions:

1. He could define a large class of subsets on which it is possible to define *curvatures*, extending the class of smooth submanifolds and convex bodies. The subsets of $\mathbb{E}^N$ belonging to this class are called *subsets with positive reach*. Basically, the main observation is that the important tool in this context is the orthogonal projection onto the studied subset. For a given convex body, this orthogonal projection is defined at every point of $\mathbb{E}^N$ and, for a smooth submanifold, it is defined on a neighborhood of it. Federer defined the class of subsets which admit locally this property, even if they are neither convex nor smooth, calling them *subsets of positive reach*.
2. He realized that this new theory could be considered as a particular (signed) measure theory on $\mathbb{E}^N$.

18.1 Subsets of Positive Reach (Federer, 1958)

The subsets of $\mathbb{E}^N$ we consider in this chapter will be systematically measurable for the standard Lebesgue measure. Following [42], we have defined in Chap. 4.1 the notion of *reach* of a subset of $\mathbb{E}^N$. We have seen that the *reach* of a convex subset is $+\infty$ and it is strictly positive for a compact smooth submanifold embedded in $\mathbb{E}^N$. In general, the reach of a subset may be strictly positive, even if the subset is neither convex nor smooth. That is why Federer gave the following definition.

Definition 42. A subset A of $\mathbb{E}^N$ has positive reach if reach$(A) > 0$ (Fig. 18.1).

Subsets with positive reach are "not too far" from smooth submanifolds and have differential properties close to them, summarized in Theorem 55.

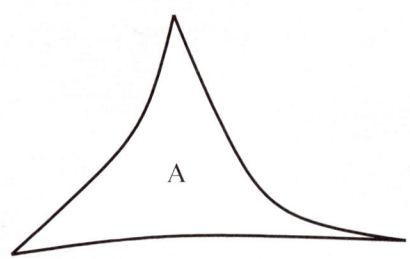

Fig. 18.1 The subset A of $\mathbb{E}^2$ has positive reach. It is nonconvex and its boundary is not a smooth regular curve

Fig. 18.2 The case of a "plain" subset A in $\mathbb{E}^2$

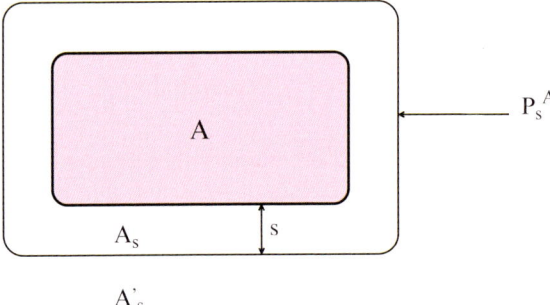

Notations. Following Federer [42], we put (Figs. 18.2 and 18.3)

$$
\begin{aligned}
A_s &= \{m : \ \mathrm{d}_A(m) \le s\}, \\
A_s' &= \{m : \ \mathrm{d}_A(m) \ge s\}, \\
P_s^A &= \{m : \ \mathrm{d}_A(m) = s\}.
\end{aligned}
\tag{18.1}
$$

Here are the main properties of the reach of a subset A. We have already seen in Chap. 4.1 that the distance function d_A to A is continuous, and the orthogonal projection map onto A is also continuous when A is compact. We improve these results in the following theorem.

Theorem 55. *Let A be a subset of $\mathbb{E}^N$:*

1. The map pr_A is continuous on U_A.
2. Moreover, if

- *m and m' lie in U_A,*
- *$\mathrm{d}_A(m) \le r, \mathrm{d}_A(m') \le r$,*
- *$\mathrm{reach}(A, \mathrm{pr}_A(m)) \ge q, \mathrm{reach}(A, \mathrm{pr}_A(m')) \ge q$,*

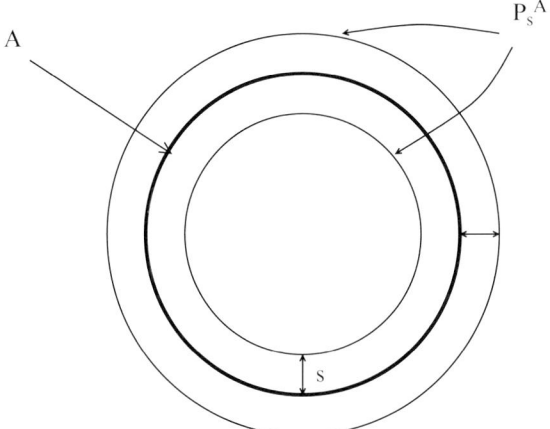

Fig. 18.3 The case of a curve A in $\mathbb{E}^2$

then

$$d(\mathrm{pr}_A(m), \mathrm{pr}_A(m')) \leq \frac{q}{q-r} d(m, m').$$

3. The subset A has positive reach $r > 0$ if and only if the function

$$d_A : \{m \in \mathbb{E}^N : 0 < d_A(m) < r\} \to \mathbb{R}$$

is C^1.

4. If A has a positive reach $r > 0$, then, for all s, t such that $0 < s < t < r$:

(a) grad d_A is Lipschitz on the open set $\{m \in \mathbb{E}^N : s < d_A(m) < t\}$.
(b) The map pr_A is Lipschitz on the set $\{m \in \mathbb{E}^N : d_A(m) \leq s\}$.
(c) In particular, the hypersurface P_s^A is a C^1-submanifold with a Lipschitz (unit) normal vector field.
(d) P_s^A admits almost everywhere a second fundamental form.

The proof is left to the reader (see [37, 42]). Let us only make precise the sense of (4-a) and (4-b). Usually, in the theory of Riemannian submanifolds, we assume that "everything is smooth." This means that we can differentiate the immersion as far as we want and, in particular, we can take the first derivative of tangent (or normal) vector fields, which gives the second fundamental tensors and the curvature tensors.

However, in our context, we do not need so much differentiability. In fact, consider for instance a hypersurface which is only C^1. Its normal vector field ξ is only continuous. Suppose moreover that ξ is Lipschitz. Then, it is almost everywhere differentiable and one can define the second fundamental form h of the hypersurface almost everywhere by taking the tangent component of $\tilde{\nabla}\xi$. In this case, h (although defined almost everywhere) is continuous, integrable, and still symmetric. This is exactly what we need to prove the *Steiner* formula for subsets with positive reach.

Remark Any compact subset of positive reach of $\mathbb{E}^N$ has finite homology. This can be seen as follows. For s small enough, we have seen that P_s is a compact C^1-submanifold and then it has finite homology. If ξ is the inward unit normal of P_s, the map

$$\eta_\xi : P_s \to A,$$

defined by

$$\forall p \in P_s, \eta_\xi(p) = p + s\xi_p,$$

sends P_s^A onto A. Since η_ξ is continuous, A has finite homology.

18.2 The Steiner Formula

The tubular neighborhood of subsets with positive reach has properties analogous to those of convex subsets or smooth submanifolds.

Theorem 56. *Let A be a subset with* positive reach r *in $\mathbb{E}^N$. Then, the volume* $\mathrm{Vol}_N(A_\varepsilon)$ *of a tubular neighborhood of A of radius $\varepsilon < r$ is a polynomial of degree less than or equal to N in ε:*

$$\mathrm{Vol}_N(A_\varepsilon) = \sum_{k=0}^{N} \Phi_k(A)\varepsilon^k. \tag{18.2}$$

Sketch of proof of Theorem 56 The crucial point is to note that, although A is not a smooth submanifold, P_s^A (with $s \leq r$) is a C^1-hypersurface of $\mathbb{E}^N$ with Lipschitz unit normal vector field, and the tube formula of Weyl (17.12) can be applied to $A'_s, s \leq r$. The proof is a direct consequence of the two following lemmas. Fix s_0 such that $\varepsilon \leq s_0 \leq r$ (Fig. 18.4). $\square$

Lemma 7.
$$\mathrm{Vol}_N(\{m \in \mathbb{E}^N : m \in (A'_{s_0})_\mu \text{ and } \mathrm{pr}_{A'_{s_0}}(m) \in P_{s_0}^A\}) \tag{18.3}$$

is polynomial in μ.

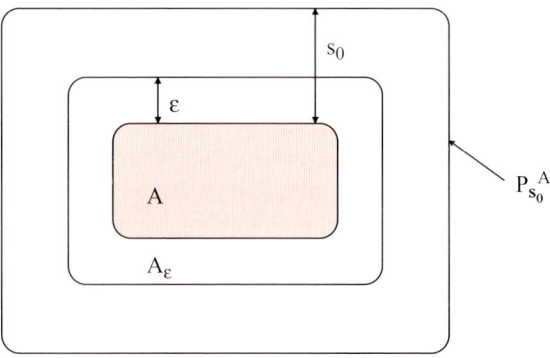

Fig. 18.4 A, A_ε, and $P_{s_0}^A$

Proof of Lemma 7 Let

$$B_\mu = \{m \in \mathbb{E}^N : m \in (A'_{s_0})_\mu \text{ and } \mathrm{pr}_{A'_{s_0}}(m) \in P^A_{s_0}\}.$$

If x is a parametrization of $P^A_{s_0}$, we define $P^A_{s_{0_t}}$ as the hypersurface generated by the parametrization $x + t\xi$, where ξ is the inward normal vector field of P_{s_0}, continuous almost everywhere. Then

$$
\begin{aligned}
\mathrm{Vol}_N(B_\mu) &= \int_0^\mu \mathrm{Vol}_{N-1}(P^A_{s_{0_t}})\,dt \\
&= \sum_{k=0}^N \int_0^\mu \Xi_k(\xi)\,dv_{P^A_{s_{0_t}}}\,t^k dt \\
&= \sum_{k=1}^N \mathbf{C}(k,n)\mathbb{M}_{k-1}(P^A_{s_0})\mu^k,
\end{aligned}
$$

where $\mathbf{C}(k,n)$ denotes a positive constant depending on the dimensions, $\Xi_k(\xi)$ denotes as before the kth-elementary symmetric functions of the principal curvatures $P^A_{s_0}$ (defined almost everywhere), and $\mathbb{M}_{k-1}(P^A_{s_0})$ denotes the global $(k-1)$-mean curvature of $P^A_{s_0}$. $\square$

Lemma 8.

$$\mathrm{Vol}_N(A_\varepsilon) = \mathrm{Vol}_N(A_{s_0}) - \mathrm{Vol}_N(B_{(s_0-\varepsilon)}). \tag{18.4}$$

Lemma 8 is obvious, using the additivity of the N-volume.

The right-hand side of (18.4) being clearly a polynomial of degree N in ε, the proof of Theorem 56 is complete. In particular, one has

$$\Phi_N(A) = \mathbb{M}_{N-1}(P^A_s). \tag{18.5}$$

Corollary 13.

1. One has

$$\Phi_i(A_s) = \sum_{j=i}^N C^i_j s^{j-i} \Phi_j(A), \ 0 \le i \le N. \tag{18.6}$$

2. In particular,

$$\Phi_N(A_s) = \Phi_N(A). \tag{18.7}$$

Proof of Corollary 13 Let u and s be such that $0 \le u < s - r$, where r denotes as before the reach of A. One has

$$\mathrm{Vol}_N((A_s)_u) = \mathrm{Vol}_N(A_{s+u}).$$

Then,

$$\sum_{k=0}^N \Phi_k(A_s)u^k = \sum_{k=0}^N \Phi_k(A)(u+s)^k,$$

from which we deduce (18.6). In particular, (18.7) is satisfied.

18.3 Curvature Measures

Federer proved a better result, since *he could localize* the *Steiner* formula, interpreting the curvatures as *Radon measures* on $\mathbb{E}^N$ (see Definition 16). In other words, the Φ_ks can be interpreted as *Radon measures.*

Theorem 57. *Let A be a subset with positive reach r in $\mathbb{E}^N$:*

1. *For every Borel subset Q (with compact closure) of $\mathbb{E}^N$, the N-volume of the subset*

$$A_\varepsilon^Q = \{m \in \mathbb{E}^N : d_A(m) \leq \varepsilon \text{ and } \mathrm{pr}_A(m) \in Q\} \tag{18.8}$$

is a polynomial in ε:

$$\mathrm{Vol}_N(A_\varepsilon^Q) = \sum_{k=0}^{N} \Phi_k(A,Q)\varepsilon^k. \tag{18.9}$$

2. *For every k, $\Phi_k(A,.)$ is a Radon measure.*
3. *Moreover, Φ satisfies the additivity property: if A, B, $A \cup B$, and $A \cap B$ have positive reach, then*

$$\Phi_k(A \cup B,.) = \Phi_k(A,.) + \Phi_k(B,.) - \Phi_k(A \cap B,.). \tag{18.10}$$

Consequently, we have the following result.

Corollary 14. *Let A be a subset of $\mathbb{E}^N$ with positive reach. For every k, $\Phi_k(A,.)$ defines a Radon measure.*

Sketch of proof of Theorem 57 Items (1) and (2) can be proved like Theorem 56, considering for a fixed Q the subset

$$\{m \in \mathbb{E}^N \text{ such that } d_{A_s}(m) \leq r \text{ and } \mathrm{pr}_{A_s}(m) \in Q\}. \quad \square$$

Definition 43. Let A be a subset of $\mathbb{E}^N$ with positive reach. For every k, $\Phi_k(A,.)$ is called the kth-curvature measure of A.

Remark To simplify the notation, we denote by the same symbol Φ_k the kth-curvature and the kth-curvature measure. In particular, with this notation, $\Phi_k(A) = \Phi_k(A,A)$.

18.4 The Euler Characteristic

The last term $\Phi_N(A)$ has a topological interpretation, extending the smooth case. In fact, we have seen in Sect. 18.1 that any compact subset with positive reach has finite homology. Then, one can consider its Euler characteristic. In the smooth case, we have used the Gauss–Bonnet theorem to relate curvature and topology. Here, the lack of differentiability leads us to use another approach. Following Federer [42], we relate the Euler characteristic of A to the degree of a suitable map. The final result is identical to the one obtained for smooth submanifolds.

Theorem 58. *Let A be a compact subset of $\mathbb{E}^N$ with positive reach. Then*

$$\Phi_N(A) = b_N \chi(A).$$

Sketch of proof of Theorem 58 Let r be the reach of A and $s < r$. Since A is a deformation retract of A_s,

$$\chi(A) = \chi(A_s).$$

To evaluate $\chi(A_s)$, we consider the map $f = d_A^2$, which is smooth in a neighborhood of A_s (see Theorem 12). The vector field

$$(\operatorname{grad} f)_{|A_s} : A_s \to \mathbb{E}^N$$

is continuous and its restriction to P_s is clearly nowhere null. Consequently, the (topological) degree of the map

$$(\operatorname{grad} f)_{|A_s} : (A_s, P_s) \to (\mathbb{E}^N, \mathbb{E}^N \setminus 0)$$

equals the degree of the map

$$(\operatorname{grad} f)_{|P_s} : P_s \to (\mathbb{E}^N \setminus 0).$$

On the other hand, since $(\operatorname{grad} f)_{|P_s}$ is proportional to the normal vector field $\overrightarrow{n}$ of P_s, $(\operatorname{grad} f)_{P_s}$ and the Gauss map G of P_s are homotopic. Hence, they have the same (topological) degree. Now, classically, the degree of G (which is – up to a constant – the Euler characteristic of P_s) is related to the integral of the second fundamental form of P_s:

$$b_N \deg G = \mathcal{M}_{N-1}(P_s).$$

Moreover, using (18.5),

$$\mathcal{M}_{N-1}(P_s) = \Phi_N(A).$$

Finally,

$$b_N \chi(A) = b_N \chi(A_s) = \deg (\operatorname{grad}) f_{|P_s} = \deg G = \mathcal{M}_{N-1}(P_s) = \Phi_N(A). \quad \square$$

Remark At this point, it becomes clear that we need a precise study of a generalized *Morse theory* [61], [64], which can be applied to functions which are not C^2. This is not trivial since *Morse theory* is based on the study of the Hessian of functions. This has been done by Banchoff [8] for polyhedra and by Fu [46] for subsets with positive reach.

18.5 The Problem of Continuity of the Φ_k

In Chap. 17, we mentioned the noncontinuity of the Φ_k for smooth submanifolds in general. A fortiori, continuity is not satisfied for subsets with positive reach. However, there are certain partial continuity results which must be mentioned. The reader can easily be convinced that these results are generalization of Theorem 52. Once again, the proof is based upon a precise study of the behavior of the distance function and the projection map.

Proposition 14. *Let $\varepsilon > 0$. Let A be a subset of $\mathbb{E}^N$ and let $(A_q)_{q \in \mathbb{N}}$ be a sequence of compact subsets of $\mathbb{E}^N$ with positive reach $\geq \varepsilon$, such that the corresponding distance functions d_{A_q} converge to d_A uniformly on every compact subset of A_ε. Then:*

1. A is compact with positive reach $\geq \varepsilon$.
2. Moreover,

$$\lim_{q \to \infty} \mathrm{pr}_{A_q} = \mathrm{pr}_A$$

uniformly on every compact subset of A_ε.

Sketch of proof of Proposition 14 The proof of compactness of A is trivial. To prove that A has positive reach $\geq \varepsilon$, we use Theorem 55 and prove that the function d_A is C^1 on $(A_\varepsilon \setminus A)$. Consider any compact C in A_ε. Let m be a point in $(A_\varepsilon \setminus A)$. We need to prove that d_A is C^1 at m. We put $d_A(m) = \alpha < \varepsilon$. Let U be an open neighborhood of m such that $\bar{U} \subset A_\varepsilon$ and let K be a compact subset of $\mathbb{E}^N$ such that

$$m \in K \subset U$$

and

$$\forall n \in K, d(m, n) < \frac{\varepsilon - \alpha}{3}.$$

Since the sequence of functions d_{A_q} converges to d_A uniformly on every compact subset of A_ε, there exists a constant $\mathbf{c}$ such that, if $q \geq \mathbf{c}$, then

$$d_A(A_q) < \frac{\varepsilon - \alpha}{3}.$$

This implies in particular that if we take

$$r = \frac{2\varepsilon}{3} + \frac{\alpha}{3} < \varepsilon,$$

we have

$$d_{A_q}(n) \leq r, \forall q > \mathbf{c}, \forall n \in K.$$

Since each d_{A_q} is continuous and the convergence is uniform, the limit is continuous, in particular at the point m. Hence, d_A is continuous at m. Moreover, we have much more than that:

- By Theorem 12, each term $\mathrm{d}^2_{A_q}$ is continuously differentiable on K. Moreover, the sequence $\mathrm{d}^2_{A_q}$ converges uniformly to d^2_A on K.
- The sequence grad $\mathrm{d}^2_{A_q}$ satisfies

$$\mathrm{grad}\, \mathrm{d}^2_{A_q}(m) = 2(m - \mathrm{pr}_{A_q}(m))$$

for every q. By Theorem 55, we know that

$$d(\mathrm{pr}_{A_q}(m), \mathrm{pr}_{A_q}(n)) = \frac{\varepsilon}{\varepsilon - r} d(m,n).$$

We deduce that

$$d(\mathrm{grad}\, \mathrm{d}^2_{A_q}(m), \mathrm{grad}\, \mathrm{d}^2_{A_q}(n)) \leq (2 + \frac{\varepsilon}{\varepsilon - r}) d(m,n).$$

- Then, the sequence $\mathrm{d}^2_{A_q}$ is equiuniformly differentiable on K. This implies that d^2_A is uniformly differentiable on K and that the sequence grad $\mathrm{d}^2_{A_q}$ converges to grad d^2_A uniformly on K.

This implies in particular that grad d_A is continuous on K and then d_A is C^1. Consequently, the reach of A is at least ε. Using the definition of weak convergence of measures (Definition 17), we can now state the following theorem. $\square$

Theorem 59.

1. *Let $\varepsilon > 0$. Let $\mathcal{R}^\varepsilon$ be the class of subsets A of $\mathbb{E}^N$ whose reach is greater than ε, endowed with the Hausdorff distance. Then for all k, the map*

$$A \rightarrow \Phi_k(A,.)$$

is continuous on $\mathcal{R}^\varepsilon$ for the weak convergence of Radon measures.
2. *If A is a subset of $\mathbb{E}^N$ with positive reach, then for every k, $\Phi_k(A_s,.)$ converges weakly to $\Phi_k(A,.)$ when s tends to 0.*

Sketch of proof of Theorem 59 We shall only prove that for all k, $(\Phi_k(A_q))_{q \in \mathbb{N}}$ converges to $\Phi_k(A)$. A slight modification of the proof gives the complete theorem. We use Proposition 14: since $\lim_{q \to \infty} A_q = A$ for the Hausdorff distance,

$$\lim_{q \to \infty} \mathrm{d}_{A_q} = \mathrm{d}_A, \lim_{q \to \infty} \chi_{A_q} = \chi_A, \lim_{q \to \infty} \mathrm{pr}_{A_q} = \mathrm{pr}_A,$$

uniformly on every compact of A_ε (χ_A denotes here the characteristic function of A), and A has positive reach $\geq \varepsilon$. Note that for a sufficiently small r,

$$\lim_{q \to \infty} \int \chi_{A_{q\varepsilon}} = \int \chi_{A_\varepsilon}$$

and consequently

$$\lim_{q \to \infty} \mathrm{Vol}_N(A_{q_r}) = \mathrm{Vol}_N(A_r).$$

Using (18.2), we get

$$\mathrm{Vol}_N(A_{q_r}) = \sum_{k=0}^{N} \Phi_k(A_q) r^k, \text{ and } \mathrm{Vol}_N(A_r) = \sum_{k=0}^{N} \Phi_k(A) r^k.$$

We deduce that, for every $k \in \{0, ..., N\}$,

$$\lim_{q \to \infty} \Phi_{A_q} = \Phi_A. \quad \square$$

18.6 The Transversal Integrals

Federer proved a kinematic formula, which allows us to relate the coefficients Φ_k to each other. This extends the kinematic formula of Chern [30] (see Theorem 53). We still denote by $G = SO(N)\alpha\mathbb{E}^N$ the group of orientation-preserving Euclidean motions of $\mathbb{E}^N$ and by $AG(N,k)$ the (oriented) affine Grassmannian of k-planes in $\mathbb{E}^N$ (see Sect. 10.7).

Theorem 60. *Let A be a subset of $\mathbb{E}^N$ with positive reach. Then for every k,*

$$\Phi_k(A) = C_N^k \frac{b_N}{b_{N-k}\gamma_{N,k}} \int_{AG(N,k)} \chi(A \cap P^k) dP^k. \tag{18.11}$$

Remarks In fact, this formula is a particular case of the more general kinematic formula, which we mentioned in the smooth case. The result is the same, in a larger context.

Theorem 61. *Let A and B be two subsets of $\mathbb{E}^N$ with positive reach, B being compact. Then for every k, there exists a sequence of constants $c_k(i,N)$ depending only on the dimension N of the ambient space, such that*

$$\int_G \Phi_k(A \cap g(B)) = \sum_{0 \le i \le k} c_k(i,N)\Phi_i(A)\Phi_{k-i}(B). \tag{18.12}$$

Theorem 61 has sense since Federer [43] proved that, if A and B have positive reach, then $A \cap g(B)$ has positive reach for almost every g. So, the left-hand side of (18.12) is well defined. Now, using the fact that the Euler characteristic satisfies

$$\chi(A \cup B) = \chi(A) + \chi(B) - \chi(A \cap B), \tag{18.13}$$

we get a second proof of Theorem 57, item (3).

Part VII
The Theory of Normal Cycles

Chapter 19
Invariant Forms

The geometric measures of suitable subsets of $\mathbb{E}^N$ (called *geometric subsets* in Chap. 20) are obtained by integration of particular differential forms of $\mathbb{E}^N \times \mathbb{S}^{N-1}$. The goal of this chapter is to define and study these forms.

19.1 Invariant Forms on $\mathbb{E}^N \times \mathbb{E}^N$

We begin with an *algebraic point of view*. Consider the group $SO(N)$ of rotations of $\mathbb{E}^N$. The action of $SO(N)$ on $\mathbb{E}^N$ can be extended in a natural way to $\mathbb{E}^N \times \mathbb{E}^N$: for any $a \in SO(N)$, we put

$$a(x,y) = (a(x), a(y)).$$

Let $\Lambda^*(\mathbb{E}^N \times \mathbb{E}^N)$ be the exterior algebra of forms on $\mathbb{E}^N \times \mathbb{E}^N$. The group $SO(N)$ acts on $\Lambda^*(\mathbb{E}^N \times \mathbb{E}^N)$ as follows. Let $a \in SO(N)$ and $\omega \in \Lambda^p(\mathbb{E}^N \times \mathbb{E}^N)$. For every $(x_k, y_k) \in \mathbb{E}^N \times \mathbb{E}^N, 1 \leq k \leq p$,

$$(a^*\omega)((x_1, y_1), \ldots, (x_p, y_p)) = \omega((a(x_1), a(y_1)), \ldots, (a(x_p), a(y_p))). \qquad (19.1)$$

We say that a form ω on $\mathbb{E}^N \times \mathbb{E}^N$ is *invariant under the diagonal action* of $SO(N)$, or simply *under rotations*, if for all $a \in SO(N)$

$$a^*\omega = \omega. \qquad (19.2)$$

This section gives an explicit expression of the forms invariant under rotations. Let $(e_1, e_2, \ldots, e_N)$ be the standard orthonormal frame of $\mathbb{E}^N$. Let us introduce the corresponding orthonormal frame of $\mathbb{E}^N \times \mathbb{E}^N$

$$\varepsilon_1 = \begin{pmatrix} e_1 \\ 0 \end{pmatrix}, \varepsilon_2 = \begin{pmatrix} e_2 \\ 0 \end{pmatrix}, \ldots, \varepsilon_N = \begin{pmatrix} e_N \\ 0 \end{pmatrix},$$

$$\widetilde{\varepsilon}_1 = \begin{pmatrix} 0 \\ e_1 \end{pmatrix}, \widetilde{\varepsilon}_2 = \begin{pmatrix} 0 \\ e_2 \end{pmatrix}, \ldots, \widetilde{\varepsilon}_N = \begin{pmatrix} 0 \\ e_N \end{pmatrix}.$$

Identifying the Euclidean space $\mathbb{E}^N$ with its dual (and consequently the k-vectors with k-forms), we define:

- The symplectic 2-form

$$\Omega = \sum_k \varepsilon_k \wedge \widetilde{\varepsilon}_k. \tag{19.3}$$

- For every $t \in \mathbb{R}$, the N-form

$$\mathcal{V} = (\varepsilon_1 + t\widetilde{\varepsilon}_1) \wedge \ldots \wedge (\varepsilon_N + t\widetilde{\varepsilon}_N). \tag{19.4}$$

This last form is polynomial in the variable t. Let $\mathcal{V}_k$ be the coefficient of t^k. If one decomposes $\Lambda^N(\mathbb{E}^N \times \mathbb{E}^N)$ as

$$\Lambda^N(\mathbb{E}^N \times \mathbb{E}^N) = \oplus_{k=0}^N \Lambda(k, N-k), \tag{19.5}$$

with

$$\Lambda(k, N-k) = \Lambda^k(\mathbb{E}^N \oplus 0) \wedge \Lambda^{N-k}(0 \oplus \mathbb{E}^N), \tag{19.6}$$

one deduces a decomposition of the form $\mathcal{V}$ as

$$\mathcal{V} = \oplus_{k=0}^N \mathcal{V}_k, \text{ with } \mathcal{V}_k \in \Lambda(k, N-k).$$

Trivially, one has

$$\mathcal{V}_k = \sum_\pi (-1)^\pi \varepsilon_{\pi(1)} \wedge \ldots \wedge \varepsilon_{\pi(k)} \wedge \widetilde{\varepsilon}_{\pi(k+1)} \wedge \ldots \wedge \widetilde{\varepsilon}_{\pi(N)}. \tag{19.7}$$

Theorem 62. *The exterior algebra of forms on $\mathbb{E}^N \times \mathbb{E}^N$ invariant under the diagonal action of $SO(N)$ is generated by the symplectic form Ω and the N-forms $\mathcal{V}_k, 0 \le k \le N$.*

We leave the proof of this purely algebraic result to the reader (see [49, 50] for instance).

Remark It is crucial for the following to note that these forms have the same expression independently of the orthonormal frame with which they are defined, since they are invariant under the group of rigid motion.

19.2 Invariant Differential Forms on $\mathbb{E}^N \times \mathbb{S}^{N-1}$

Let us continue with a *differential point of view*. For further use, we now restrict our attention to invariant $(N-1)$-form, defined on $\mathbb{E}^N \times \mathbb{S}^{N-1}$, modifying a little the notion of invariance. Denote by **D** the *Poincaré group* of rigid motions of $\mathbb{E}^N$. It is well known that **D** is the semidirect product of the group of translations (identified to $\mathbb{E}^N$) and the group $SO(N)$ of rotations:

$$\mathbf{D} = \mathbb{E}^N \propto SO(N). \qquad (19.8)$$

The action of $\mathbf{D}$ on $\mathbb{E}^N$ can be extended in a natural way to $\mathbb{E}^N \times \mathbb{S}^{N-1}$ by putting

$$\phi(x, \eta) = ((t, a)x, a\eta), \forall x \in \mathbb{E}^N, \forall \eta \in \mathbb{S}^{N-1}$$

for all $\phi = (t, a) \in \mathbf{D}$.

Consider the space $\mathcal{D}^{N-1}(\mathbb{E}^N \times \mathbb{S}^{N-1})$ of *differential* $(N-1)$-forms on $\mathbb{E}^N \times \mathbb{S}^{N-1}$. The group $\mathbf{D}$ acts on this space as follows. Let $\phi \in \mathbf{D}$. $\forall (X_k, \eta_k) \in T(\mathbb{E}^N \times \mathbb{S}^{N-1}), 1 \leq k \leq N-1$,

$$\phi^*(\omega)((X_1, \eta_1), \ldots, (X_{N-1}, \eta_{N-1})) = \omega(d\phi(X_1, \eta_1), \ldots, d\phi(X_{N-1}, \eta_{N-1})). \qquad (19.9)$$

Definition 44. A differential $(N-1)$-form ω on $\mathbb{E}^N \times \mathbb{S}^{N-1}$ is *invariant under rigid motion* if, for all $\phi \in \mathbf{D}$,

$$\phi^* \omega = \omega.$$

We now give an explicit expression for the invariant differential $(N-1)$-forms. Let (m, ξ) be a point of $\mathbb{E}^N \times \mathbb{S}^{N-1}$. Consider an orthonormal frame of $\mathbb{E}^N$ whose N^{th}-vector is ξ:

$$(e_1, e_2, \ldots, e_{N-1}, e_N = \xi).$$

(The choice of putting ξ at the last entry is totally arbitrary and will appear to be convenient later.)

Then, an orthonormal frame of $T_{(m, \xi)} \mathbb{E}^N \times \mathbb{S}^{N-1}$ is given by the vectors

$$\varepsilon_1 = \begin{pmatrix} e_1 \\ 0 \end{pmatrix}, \varepsilon_2 = \begin{pmatrix} e_2 \\ 0 \end{pmatrix}, \ldots, \varepsilon_N = \begin{pmatrix} e_N = \xi \\ 0 \end{pmatrix},$$

$$\tilde{\varepsilon}_1 = \begin{pmatrix} 0 \\ e_1 \end{pmatrix}, \tilde{\varepsilon}_2 = \begin{pmatrix} 0 \\ e_2 \end{pmatrix}, \ldots, \tilde{\varepsilon}_{N-1} = \begin{pmatrix} 0 \\ e_{N-1} \end{pmatrix}.$$

Identifying the Euclidean space $\mathbb{E}^N$ with its dual (and consequently the $(N-1)$-vectors with $(N-1)$-forms), we can define the following $(N-1)$-form:

$$\mathcal{V}(t) = (\varepsilon_1 + t\tilde{\varepsilon}_1) \wedge \ldots \wedge (\varepsilon_{N-1} + t\tilde{\varepsilon}_{N-1})_{(m, \xi)}. \qquad (19.10)$$

This defines a differential $(N-1)$-form on $ST\mathbb{E}^N = \mathbb{E}^N \times \mathbb{S}^{N-1}$, which can be considered as a polynomial in the variable t. Let $\mathcal{W}_k$ be the coefficient of t^k. Then, $\mathcal{W}_k$ is a differential form which does not depend on the orthonormal frame $(e_1, \ldots, e_N = \xi)$.

Definition 45. For $0 \leq k \leq N-1$, the $(N-1)$-differential form $\mathcal{W}_k$, defined on $\mathbb{E}^N \times \mathbb{S}^{N-1}$ as the coefficient of t^k in $\mathcal{V}(t)$, is called the kth Lipschitz–Killing curvature form.

Trivially, one has

$$\mathcal{W}_k = \sum_{\pi} (-1)^{\pi} \varepsilon_{\pi(1)} \wedge \dots \wedge \varepsilon_{\pi(k)} \wedge \widetilde{\varepsilon}_{\pi(k+1)} \wedge \dots \wedge \widetilde{\varepsilon}_{\pi(n-1)}. \tag{19.11}$$

Roughly speaking, these forms span the C^{∞}-module of differential $(N-1)$-forms on $\mathbb{E}^N \times \mathbb{S}^{N-1}$ invariant under $\mathbf{D}$. More precisely, one has the following theorem.

Theorem 63. *The vector space of invariant differential $(N-1)$-forms on $\mathbb{E}^N \times \mathbb{S}^{N-1}$ has dimension N if N is even and $N+1$ if N is odd. It is spanned by the forms $(\mathcal{W}_k)_{(k=0,\dots,(N-1))}$ and, if N is odd, by $\Omega^{\frac{N-1}{2}}$ (where Ω denotes the standard symplectic form of $\mathbb{E}^N \times \mathbb{E}^N$ restricted to $\mathbb{E}^N \times \mathbb{S}^{N-1}$).*

19.3 Examples in Low Dimensions

- Suppose in particular that $N=2$. Then, the space of invariant 1-forms of $\mathbb{E}^2 \times \mathbb{S}^1$ has dimension 2. At a point $(p, \xi = e_2)$, it is spanned by the two 1-forms

$$\mathcal{W}_0 = \varepsilon_1,$$
$$\mathcal{W}_1 = \widetilde{\varepsilon}_1.$$

- Suppose now that $N=3$. Then, the space of invariant 2-forms of $\mathbb{E}^3 \times \mathbb{S}^2$ has dimension 4. At a point $(p, \xi = e_3)$, it is spanned by the four 2-forms

$$\mathcal{W}_0 = \varepsilon_1 \wedge \varepsilon_2,$$
$$\mathcal{W}_1 = \varepsilon_1 \wedge \widetilde{\varepsilon}_2 + \widetilde{\varepsilon}_1 \wedge \varepsilon_2,$$
$$\mathcal{W}_2 = \widetilde{\varepsilon}_1 \wedge \widetilde{\varepsilon}_2,$$
$$\mathcal{W}_3 = \Omega_{|\mathbb{E}^N \times \mathbb{S}^{N-1}} = \varepsilon_1 \wedge \widetilde{\varepsilon}_1 + \varepsilon_2 \wedge \widetilde{\varepsilon}_2.$$

Remark Once again, note that these forms have the same expression independently of the orthonormal frame with which they are defined, since they are invariant under the group of rigid motion.

Chapter 20
The Normal Cycle

This chapter is the heart of the book. As we wrote in the introduction, a major advance in the construction of generalized curvatures was accomplished by Wintgen and Zähle: these authors wanted to include simple subsets like polyhedra in the theory. They replaced the measures of Federer by *differential forms on the unit sphere bundle $ST\mathbb{E}^N$ invariant under rigid motions*, which can be integrated over any current (defined in Chap. 19). Then, they associated to suitable subsets $\mathcal{A}$ of $\mathbb{E}^N$ an $(N-1)$-current on $ST\mathbb{E}^N$ called *the normal cycle $\mathbf{N}(A)$ of $\mathcal{A}$*. This current generalizes the unit normal bundle of submanifolds and has two basic properties:

1. $\mathbf{N}$ is additive:
$$\mathbf{N}(\mathcal{A}\cup\mathcal{B}) = \mathbf{N}(\mathcal{A}) + \mathbf{N}(\mathcal{B}) - \mathbf{N}(\mathcal{A}\cap\mathcal{B}) \qquad (20.1)$$
   (when it makes sense).
2. $\mathbf{N}(\mathcal{A})$ contains the geometry of $\mathcal{A}$. More precisely, by integrating on $\mathbf{N}(\mathcal{A})$ the invariant differential forms, one obtains the *Quermassintegrale* defined on $\mathcal{A}$ when $\mathcal{A}$ is the boundary of a convex domain and the usual global *Lipschitz–Killing curvatures* when $\mathcal{A}$ is a smooth compact submanifold of $\mathbb{E}^N$.

These properties are natural. Moreover, it becomes clear that, to study the geometry of a complicated object, it is enough to decompose the object into simple parts and apply the fundamental property (20.1). In particular, it becomes possible to study the curvatures of polyhedra, which are the union of elementary *convex* polyhedra, since the normal cycle associated to any convex polyhedron is easy to evaluate.

20.1 The Notion of a Normal Cycle

Up to now, there is no general method to determine the class of subsets of $\mathbb{E}^N$ which "admit a normal cycle." We shall therefore adopt an empirical way, defining a "normal cycle" as a (closed) $(N-1)$-integral current of $\mathbb{E}^N \times \mathbb{S}^{N-1}$ associated to "simple" objets of $\mathbb{E}^N$. More precisely, we shall define the *normal cycle* of:

- A smooth submanifold of $\mathbb{E}^N$
- A subset of positive reach of $\mathbb{E}^N$
- A simplex and a polyhedron of $\mathbb{E}^N$

20.1.1 Normal Cycle of a Smooth Submanifold

Let M^n be a (compact smooth) submanifold of $\mathbb{E}^N$. The *normal cycle* of M^n is the cycle canonically associated to its unit normal bundle (denoted by $\mathbf{N}(M^n)$).

Remarks

1. By definition, the normal cycle of a smooth submanifold is the current whose support is a smooth oriented closed submanifold of $\mathbb{E}^N \times \mathbb{S}^{N-1}$ of dimension $(N-1)$, endowed with its canonical orientation.
2. For any sufficiently small ε, there is a canonical one-to-one correspondence between ∂M_ε^n and (the support of) $\mathbf{N}(M^n)$: with obvious notations, to each point $(m+\varepsilon\xi)$, one associates the point (m,ξ). The map

$$\phi : (m+\varepsilon\xi_m) \rightarrow (m,\xi)$$

induces an orientation on $\mathbf{N}(M^n)$.

20.1.2 Normal Cycle of a Subset of Positive Reach

If $\mathcal{A}$ is a subset with positive reach, we have seen in Chap. 18 that, for any sufficiently small ε, the subset $P_s^{\mathcal{A}}$ (defined in (18.1)) is an (oriented) C^1-hypersurface of $\mathbb{E}^N$ with a Lipschitz unit normal vector field. One can associate to $P_s^{\mathcal{A}}$ an $(N-1)$-cycle of $\mathbb{E}^N \times \mathbb{S}^{N-1}$, whose support is the image of the exponential map

$$\phi : (m+\varepsilon\xi_m) \rightarrow (m,\xi_m),$$

for every m in $\mathcal{A}$ and every ξ_m in the normal cone of $\mathcal{A}$ at m. It is an $(N-1)$-dimensional closed (oriented) C^1-submanifold, the orientation of $\mathbf{N}(\mathcal{A})$ being induced by ϕ. This cycle is obviously rectifiable. It is called the *normal cycle* of $\mathcal{A}$ and denoted by $\mathbf{N}(\mathcal{A})$.

One of the main problems which appears now is the fact that, if $\mathcal{A}$ and $\mathcal{B}$ are subsets of $\mathbb{E}^N$ with positive reach, $\mathcal{A} \cap \mathcal{B}$ is not a subset with positive reach in general. So, (20.1) cannot be applied to compute $\mathbf{N}(\mathcal{A} \cup \mathcal{B})$. An extensive study of this difficulty can be found in [88].

20.1.3 Normal Cycle of a Polyhedron

The main notations and definitions have been given in Chap. 6. Remember that, in general, a polyhedron of $\mathbb{E}^N$ is not a subset with positive reach (the basic counterexample is the polygon composed of two segments $PQ \cup QR$). Thus, the classical theory cannot be applied. However, every k-simplex in $\mathbb{E}^N$ is convex and therefore has positive reach. Then, any polyhedron can be decomposed into subsets of positive reach. Moreover, the intersection of two simplices composing a polyhedron is still a simplex. Consequently, it is natural to begin by defining the normal cycle of a simplex.

Definition 46. Let σ^k be an elementary k-simplex of $\mathbb{E}^N$. The normal cycle of σ^k is the (closed) current $\mathbf{N}(\sigma^k)$ of $\mathbb{E}^N \times \mathbb{S}^{N-1}$, whose support $\mathrm{spt}(\mathbf{N}(\sigma^k))$ is the (smooth) hypersurface defined by

$$\mathrm{spt}(\mathbf{N}(\sigma^k)) = \{(q, \xi) \in \mathbb{E}^N \times \mathbb{S}^{N-1}, \text{ where } q \in \sigma^k, \text{ and } \xi \in C_q\} \qquad (20.2)$$

and whose orientation is given by the outward normal.

To define the normal cycle for any polyhedron, we use the fact that it is the union of simplices and use the relation (20.1): if σ^k and σ^l are two simplices,

$$\mathbf{N}(\sigma^k \cup \sigma^l) = \mathbf{N}(\sigma^k) + \mathbf{N}(\sigma^l) - \mathbf{N}(\sigma^k \cap \sigma^l). \qquad (20.3)$$

We leave the proof to the reader that the normal cycle obtained by this process is independent of the decomposition (Fig. 20.1).

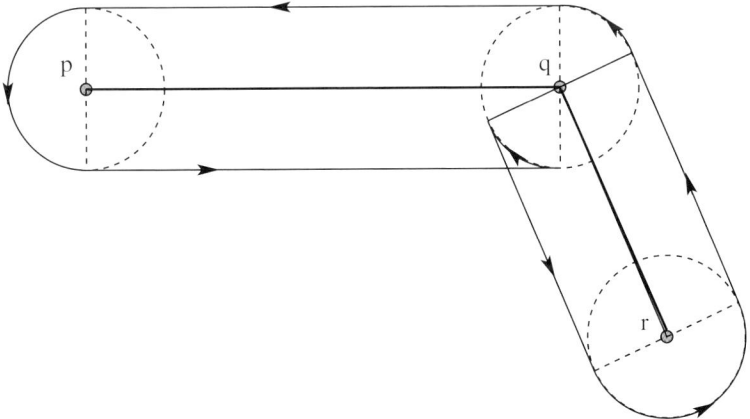

Fig. 20.1 The normal cycle of the union of two segments pq and qr is the cycle defined by the (oriented) plain curve. To compute the normal cycle of the union $\mathcal{A}$ of the two segments pq and qr in the plane, one remarks that the support of the normal cycle of pq (resp., qr) is the union of two segments and two half-circles. The intersection of pq and qr is the point q. The support of the normal cycle of q is a circle. Now, we use (20.3) by assigning the suitable orientation to each interval or arc of circle composing the support of the normal cycle of pq and qr

20.1.4 Normal Cycle of a Subanalytic Set

Fu [49] and Kashiwara and Schapira [54, Chap. 9] extended the notion of *normal cycle* to the large class of subanalytic subsets of $\mathbb{E}^N$. The reader can consult [49, 54] for details, beyond the scope of this book, and the work of Bernig [14] for more recent results.

20.2 Existence and Uniqueness of the Normal Cycle

In the previous sections, we have associated to some simple subsets of $\mathbb{E}^N$ a current called the *normal cycle* which generalizes the unit normal bundle. Following Fu [47], we now characterize the normal cycle of a compact subset of $\mathbb{E}^N$ and give a uniqueness result. Note, however, that this theorem will not claim the *existence* of such a normal cycle associated to *any* compact subset. That is why Fu gave a very general method to construct such a cycle under slight assumptions.

Let $\mathcal{A}$ be a compact subset of $\mathbb{E}^N$. Consider the function

$$i_{\mathcal{A}} : ST\mathbb{E}^N \to \mathbb{R},$$

defined by

$$i_{\mathcal{A}}(x,\xi) = \lim_{r \to 0}\lim_{s \to 0}[\chi(A \cap B(x,r) \cap \{p \text{ such that } (p-x).\xi \leq t\})|_{t=-s}^{t=+s}],$$

where χ denotes here the Euler characteristic (Figs. 20.2 and 20.3).

Note that $i_{\mathcal{A}}$ may have no sense if the Euler characteristic is taken for an object which has no finite homology. On the other hand, when $\mathcal{A}$ is a stratified set, $i_{\mathcal{A}}(x,\xi)$ is just the index of x considered as a critical point of the height function defined by

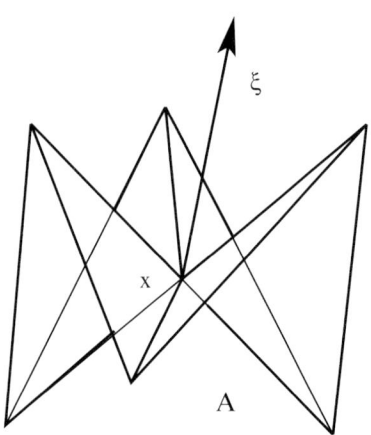

Fig. 20.2 In this example, $i_{\mathcal{A}}(x,\xi) = 0$

Fig. 20.3 In this example,
$i_{\mathcal{A}}(x,\xi) = 4$

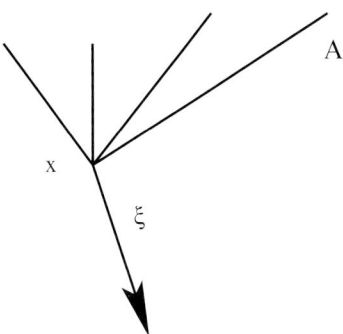

ξ. In some sense, $i_{\mathcal{A}}(x,\xi)$ can be viewed as the multiplicity of (x,ξ). In any case, if $i_{\mathcal{A}}$ exists, it is unique in a certain sense. Let us explain this last point. Remember that the normal bundle $T^{\perp}M^n$ of a smooth submanifold M^n (of $\mathbb{E}^N$) is Lagrangian with respect to the canonical symplectic structure Ω of $T\mathbb{E}^N \simeq \mathbb{E}^N \times \mathbb{E}^N$, and the unit normal bundle $ST^{\perp}M^n$ is Legendrian with respect to the canonical Liouville form α on $ST\mathbb{E}^N = \mathbb{E}^N \times \mathbb{S}^{N-1}$ (see Theorem 34). We mimic this property for integral currents in the next definition.

Definition 47. An integral $(N-1)$-current C of $\mathbb{E}^N \times \mathbb{S}^{N-1}$ is Legendrian if it satisfies

$$< C, \alpha \wedge \omega >= 0 \qquad (20.4)$$

for all $\omega \in \mathcal{D}^{N-2}(\mathbb{E}^N \times \mathbb{S}^{N-1})$.

Fu [48] proved the following theorem.

Theorem 64. *Let $\mathcal{A}$ be a compact subset of $\mathbb{E}^n$. There exists at most one closed compactly supported integral current $S_{\mathcal{A}} \in I_{N-1}(ST\mathbb{E}^N)$ such that:*

- *$S_{\mathcal{A}}$ is Legendrian.*
- *For all smooth functions φ in $ST\mathbb{E}^N$,*

$$< S_{\mathcal{A}}, \varphi(x,\xi)dv_{\mathbb{S}^{n-1}} >= \int_{\mathbb{S}^{N-1}} \sum_{x \in \mathbb{E}^n} \varphi(x,\xi)i_{\mathcal{A}}(x,\xi)dv_{\mathbb{S}^{N-1}}.$$

Following [48], we give the following result.

Definition 48. Let $\mathcal{A}$ be a compact subset of $\mathbb{E}^n$. If $i_{\mathcal{A}}$ and $S_{\mathcal{A}}$ exist, then $\mathcal{A}$ is said to be geometric. In this case, $S_{\mathcal{A}}$ is denoted by $\mathbf{N}(\mathcal{A})$. It is called the normal cycle of $\mathcal{A}$.

From the additivity of the Euler characteristic, one deduces the following important theorem.

Theorem 65. *Let $\mathcal{A}, \mathcal{B}, \mathcal{A} \cap \mathcal{B}$, and $\mathcal{A} \cup \mathcal{B}$ be four compact subsets of $\mathbb{E}^N$. If three of them are geometric, so is the fourth, and*

$$\mathbf{N}(\mathcal{A} \cup \mathcal{B}) = \mathbf{N}(\mathcal{A}) + \mathbf{N}(\mathcal{B}) - \mathbf{N}(\mathcal{A} \cap \mathcal{B}). \qquad (20.5)$$

Of course, the reader must be convinced that Definition 48 and Theorem 65 are consistent with our definition of the normal cycle of smooth submanifolds and polyhedra. We leave this verification to the reader.

20.3 A Convergence Theorem

This section deals with a general theorem of convergence of normal cycles previously defined. Fu [48] proved this result for a sequence $(P_p)_{p \in \mathbb{N}}$ of triangulated polyhedra closely inscribed in a submanifold M^n of the Euclidean space (in the sense of Definition 37) converging to M^n for the Hausdorff topology. He showed that the sequence of normal cycles of P_p converges to the normal cycle of M^n for the flat topology of currents. This result is a consequence of the compactness theorem for integral currents mentioned in Chap. 12.

Theorem 66. *Let M^n be a smooth submanifold of $\mathbb{E}^N$. Let $(P_p)_{p \in \mathbb{N}}$ be a sequence of polyhedra closely inscribed in M^n such that:*

1. The Hausdorff limit of P_p is M^n when p tends to infinity.
2. The Hausdorff limit of ∂P_p is ∂M^n when p tends to infinity.
3. The fatness of P_p is uniformly bounded from below by a nonnegative constant:

$$\forall p \geq 0, \exists c > 0 \ such \ that \ \Theta(P_p) > c > 0.$$

Then,

$$\lim_{p \to \infty} \mathbf{N}(P_p) = \mathbf{N}(M)$$

for the flat topology.

The basic tools of this result are the convergence Theorem 40 for volumes which we proved in Chap. 13 and the compactness theorem for bounded integral currents (see Theorem 36). The assumptions are those described in Theorem 40: we assume that the sequence of polyhedra closely inscribed in a smooth submanifold of $\mathbb{E}^N$ converges to it with a bounded fatness. Under these assumptions, the sequence of volumes of P_p tends to the volume of M^n. This implies in particular that the volumes of P_p are uniformly bounded.

Sections 20.3.1 and 20.3.2 give details of the proof of Theorem 66.

20.3.1 Boundness of the Mass of Normal Cycles

Now, the crucial point is to bound the sequence of the *masses* of the normal cycles $\mathbf{N}(P_p)$.

Proposition 15. *Under the assumptions of Theorem 66, the masses of the currents* $\mathbf{N}(P_p)_{p\in\mathbb{N}}$ *are uniformly bounded by a positive constant $c > 0$.*

The proof of this proposition uses the following lemma, which is a consequence of the uniform fatness of the sequence of triangulations.

Lemma 9. *Under the assumptions of Theorem 66, there exists a constant $C > 0$ such that for each $p \in \mathbb{N}$ and each vertex v of P_p, the number of simplices of P_p incident to v is less than C.*

As a consequence of this lemma and the fact that the volumes of the polyhedra P_p are uniformly bounded by positive constant C, we can deduce that, for a fixed p, the sum of the masses of the normal cycle localized over the vertices of each P_p is bounded by a positive constant C_0, independent of p. Using the same kind of argument, we show that, for each k, the masses of the normal cycles localized over the k-simplices of P_p are uniformly bounded by a positive constant C_k, independent of p. The proposition follows by using the additivity formula (20.5) for the normal cycle, from which we deduce that

$$\forall p \in \mathbb{N}, \mathbf{M}(\mathbf{N}(P_p)) \leq C_0 + \ldots + C_n$$

and Proposition 15 is proved by taking $c = C_0 + \ldots + C_n$.

20.3.2 Convergence of the Normal Cycles

Using Sect. 20.3.1, we can now show that the sequence of normal cycles $\mathbf{N}(P_p)_{p\in\mathbb{N}}$ converges when p tends to infinity. First, we apply the compactness Theorem 36 for integral currents mentioned in Chap. 12 to the sequence $\mathbf{N}(P_p)_{p\in\mathbb{N}}$. We deduce that there exists a subsequence $\mathbf{N}(P_{p_q})_{q\in\mathbb{N}}$ of $\mathbf{N}(P_p)_{p\in\mathbb{N}}$, which converges to an integral current T. Now, by construction, the support of T lies in M^n and the support of ∂T lies in ∂M^n. Then, the constancy Theorem 35 implies that

$$\lim_{q\to\infty} \mathbf{N}(P_{p_q}) = \mathbf{N}(M)$$

(in our framework, the convergence of the volumes implies that $a = 1$). Since this equality is independent of the choice of the convergent subsequence, we deduce that the sequence of normal cycles $\mathbf{N}(P_p)_{p\in\mathbb{N}}$ converges when p tends to infinity.

20.4 Approximation of Normal Cycles

This section deals with a general approximation result: we assume that a geometric subset $\mathcal{W}$ of $\mathbb{E}^N$ (in the sense of Definition 48) is close to a compact (smooth) hypersurface W and we give an approximation of the curvatures of W with respect to those of $\mathcal{W}$. Basically, we evaluate an upper bound on the flat norm of the difference of the normal cycle of the compact subset K of $\mathbb{E}^N$ whose boundary is W and the normal cycle of the geometric compact subset $\mathcal{K}$ whose boundary is $\mathcal{W}$ (see Definitions 49 and 50). This result [35] can be considered as a quantitative version of the convergence Theorem 66. Note that we do not use the compactness theorem for currents, which is a crucial tool in the proof of Theorem 66. Although we restrict our study to hypersurfaces for simplicity, the reader can adapt the results to any codimension.

We are not able to prove our result in full generality: for technical reasons, we need a restrictive condition on a "small part" of the boundary $\mathcal{W}$ of $\mathcal{K}$. We introduce the following result.

Definition 49. A subset $\mathcal{K}$ of $\mathbb{E}^n$ is weakly regular if there exists a point of $\partial \mathcal{K}$ having a neighborhood in $\mathcal{K}$ diffeomorphic to a half-space.

For instance, a codimension one polyhedron is weakly regular. This definition may seem somewhat artificial, but we need it in the proof of Theorem 67 when we apply the constancy theorem for integral currents.

The main quantity involved in the study of the couple K and $\mathcal{K}$ is the *angular deviation*, which is a direct generalization of the angular deviation defined for submanifolds "close to each other" in Definition 34. We give a precise definition.

Definition 50. Let W be a (compact smooth oriented) hypersurface of $\mathbb{E}^N$ and $\mathcal{W}$ be a geometric subset of $\mathbb{E}^N$ closely near W:

- Let $p \in \mathcal{W}$. The angular deviation between p and $\mathrm{pr}_W(p)$[1] is the maximal angle α_p between n_p and $\xi_{\mathrm{pr}_W(p)}$ when $(p, n_p) \in \mathrm{spt}\,\mathbf{N}(\mathcal{K})$ (where ξ denotes the unit normal vector field of W).
- If B is any Borel subset of $\mathcal{W}$, the deviation angle between B and $\mathrm{pr}_W(B)$ is the real number $\alpha_B = \sup_{p \in B} \alpha_p$.

We can now prove the following theorem.

Theorem 67. *If $\mathcal{W}$ is weakly regular and closely near W, then, for any (connected) Borel subset $B \subset \mathcal{W}$,*

$$\mathcal{F}(\mathbf{N}(\mathcal{K})_{|T_B \mathbb{E}^N} - \mathbf{N}(K)_{|T_{\mathrm{pr}_W(B)} \mathbb{E}^N}) \le$$
$$(\delta_B + \alpha_B)\left[\frac{2(1+|h_{\mathrm{pr}_W(B)}|)}{1 - \delta_B |h_{\mathrm{pr}_W(B)}|}\right]^{n-1} (\mathbf{M}(\mathbf{N}(\mathcal{K})_{|T_B \mathbb{E}^N}) + \mathbf{M}(\partial \mathbf{N}(\mathcal{K}_{|T_B \mathbb{E}^N}))),$$

---

[1] As before, $\mathrm{pr}_W(p)$ denotes the orthogonal projection of p onto W.

where δ_B denotes the Hausdorff distance between B and $\mathrm{pr}_W(B)$, and $\left|h_{\mathrm{pr}_W(B)}\right|$ denotes the supremum over $\mathrm{pr}_W(B)$ of the norm of the second fundamental form of W.

The rest of this chapter is devoted to the proof of Theorem 67. A detailed proof can be found in [35].

Sketch of proof of Theorem 67 By assumption, W lies in $U = U_r(W)$, where r is the reach of W. Consider the map f defined by the diagram

$$
\begin{array}{ccc}
T_U \mathbb{E}^N & \xrightarrow{\ f\ } & \mathrm{spt}\,\mathbf{N}(K) \\
\pi \downarrow & & \uparrow G \\
U & \xrightarrow[\mathrm{pr}]{} & W
\end{array}
$$

To simplify the notations, we define the $(n-1)$-current D by

$$
D = \mathbf{N}(\mathcal{K})_{|T_B \mathbb{E}^N}
$$

and E by

$$
E = \mathbf{N}(K)_{|T_{\mathrm{pr}(B)} \mathbb{E}^N}. \qquad \square
$$

Lemma 10. *One has*

$$
f_\sharp(D) = E.
$$

Proof of Lemma 10 We apply the constancy Theorem 35 to f, $\mathbf{N}(\mathcal{K})$, and $\mathbf{N}(K)$: since the support of the image by f of $\mathbf{N}(\mathcal{K})$ is included in the support of $\mathbf{N}(K)$, there exists an integer c such that

$$
f_\sharp(\mathbf{N}(\mathcal{K})) = c\mathbf{N}(K). \tag{20.6}
$$

We need to prove that $c = 1$. First of all, by a classical property of (proper) smooth maps between currents (see [43, p. 359]), one has

$$
f_\sharp(\mathbf{N}(\mathcal{K})_{|f^{-1}(A)}) = f_\sharp(\mathbf{N}(\mathcal{K}))_{|A} \tag{20.7}
$$

for every Borel subset A of $T_U \mathbb{E}^N$. Then, (20.6) and (20.7) imply that

$$
f_\sharp(\mathbf{N}(\mathcal{K})_{|f^{-1}(A)}) = c\mathbf{N}(K)_{|A}.
$$

We will apply this equality to the following subset A. By assumption, $\mathcal{W}$ contains a point having a neighborhood (in $\mathcal{K}$) diffeomorphic to a half-space. In this neighborhood, there exists a neighborhood $\mathcal{U}$ whose closure is diffeomorphic to a half-ball. The subset $\mathcal{K}$ is the union of $\mathcal{U}$ and the closure of $\mathcal{K}\backslash\mathcal{U}$. Since $\mathcal{U}$, K, and $\overline{\mathcal{K}\backslash\mathcal{U}} \cap \mathcal{U}$ are geometric, we can apply the additivity property of the normal cycles (20.5) and deduce that the normal cycle of $\mathcal{K}$ over $\mathcal{U} \cap \mathcal{W}$ is the current associated to the unit normal bundle (of dimension 1) of $\mathcal{U} \cap \mathcal{W}$. Since the restriction of pr_W to $\mathcal{U}$ is one-to-one onto $\mathrm{pr}_W(\mathcal{U})$, the restriction of f to the support of $\mathbf{N}(\mathcal{K})_{|T_\mathcal{U} \mathbb{E}^n}$ is (smooth and) one-to-one onto $N(K)_{|T_{\mathrm{pr}_W \mathcal{U}} \mathbb{E}^n}$, and

$$f_\sharp((\mathbf{N}(\mathcal{K})_{|T_\mathcal{U}\mathbb{E}^N}) = \mathbf{N}(K)_{|T_{\mathrm{pr}_W(\mathcal{U})}\mathbb{E}^n}. \tag{20.8}$$

Taking $A = T_{\mathrm{pr}_W(\mathcal{U})}\mathbb{E}^n$ in (20.7), we deduce that $c = 1$. $\square$

Now, we define a homotopy g between f and the identity. We put

$$g : T_{\tilde{B}}\mathbb{E}^n \times [0,1] \to T\mathbb{E}^n, \tag{20.9}$$

with

$$g(X_x,t) = (1-t)X_x + tf(X_x).$$

We define the n-current:

$$C = g_\sharp(D \times [0,1]).$$

The homotopy formula for currents (see Theorem 37) gives immediately

$$\partial C = f_\sharp(D) - D - g_\sharp(\partial D \times [0,1]).$$

Moreover, we have the following proposition.

Proposition 16.

$$\mathcal{F}(D - E) \le (\mathbf{M}(D) + \mathbf{M}(\partial D)) \sup_t \sup_{\mathrm{spt}D}[(|\frac{dg}{dt}|)(|dg_t|^{n-2}, |dg_t|^{n-1})].$$

Proof of Proposition 16 To evaluate the flat norm of $D - E$, we decompose $D - E$ into a sum of an $(n-1)$-current and the boundary of an n-current, by writing

$$D - E = \partial C - g_\sharp(\partial D \times [0,1]).$$

By definition of the flat norm,

$$\mathcal{F}(D - E) \le \mathbf{M}(C) + \mathbf{M}(g_\sharp(\partial D \times [0,1])).$$

To evaluate $\mathbf{M}(C)$ and $\mathbf{M}(g_\sharp(\partial D \times [0,1]))$, we use the fact that D is representable by integration. By a computation similar to [42, 4.1.9], we have

$$\mathbf{M}(C) = \mathbf{M}(g_\sharp(D \times [0,1]))$$

$$\le \int_{[0,1]} \mathbf{M}(D) \sup_{\mathrm{spt}D \times \{t\}} (|\frac{dg}{dt}|) \sup_{\mathrm{spt}D \times \{t\}} (|dg_t|^{n-1}) dt$$

and

$$\mathbf{M}(g_\sharp(\partial D \times [0,1])) \le \int_{[0,1]} \mathbf{M}(\partial D) \sup_{\mathrm{spt}D \times \{t\}} |\frac{dg}{dt}| \sup_{\mathrm{spt}D \times \{t\}} (|dg_t|^{n-2}) dt,$$

from which we deduce Proposition 16. $\square$

End of the proof of Theorem 67 Now, B satisfies

$$\delta_B |h_{\mathrm{pr}_W(B)}| < 1,$$

$$|d\mathrm{pr}_{|\tilde{B}}| \le \frac{1}{1 - \delta_B |h_{\mathrm{pr}(B)}|},$$

and

$$|dg_t|^{n-1} \le (1-t) + t\left(\frac{1 + |h_{\mathrm{pr}_W(B)}|}{1 - \delta_B |h_{\mathrm{pr}_W(B)}|}\right)^{n-1} \le \left(\frac{2(1 + |h_{\mathrm{pr}_W(B)}|)}{1 - \delta_B |h_{\mathrm{pr}_W(B)}|}\right)^{n-1},$$

from which we deduce Theorem 67. $\square$

Chapter 21
Curvature Measures of Geometric Sets

We give here a general framework to compute the kth-*generalized curvature* of a *geometric* subset of $\mathbb{E}^N$ as defined in Definition 48. We also apply the results of Chap. 20 to obtain convergence and approximation theorems on curvature measures.

21.1 Definition of Curvatures

If C is any $(N-1)$-current on $\mathbb{E}^N \times \mathbb{S}^{N-1}$, the kth-curvature $\mathcal{M}_k(C)$ of C can be defined by the formula

$$\mathcal{M}_k(C) = < C, \mathcal{W}_k > .$$

To define the curvature of any geometric (compact) subset of $\mathbb{E}^N$, we use its normal cycle.[1]

Definition 51. Let $\mathcal{A}$ be a (compact) geometric subset of $\mathbb{E}^N$. Then, the kth-curvature $\mathcal{M}_k(\mathcal{A})$ of $\mathcal{A}$ is the real number defined by the formula

$$\mathcal{M}_k(\mathcal{A}) = < \mathbf{N}(\mathcal{A}), \mathcal{W}_k > . \tag{21.1}$$

In particular cases (convex subsets, smooth submanifolds, subsets with positive reach, and polyhedra), the simple construction of the corresponding normal cycles allows us to evaluate these curvatures explicitly. We shall see later that they coincide (up to a constant depending on the dimensions) with the coefficients Φ_k of the Steiner polynomials of the studied objects.

---

[1] A complete theory should assign a series of curvatures to *any* (compact) subset of $\mathbb{E}^N$. However, we have seen that, up to now, we do not have a general theory to construct a suitable normal cycle for *any* compact subset of $\mathbb{E}^N$. A possible way to extend the theory to "all compact subsets" could be to modify *the definition* of the normal cycle given in Chap. 20.

21.1.1 The Case of Smooth Submanifolds

Let M^n be an n-dimensional (smooth) submanifold of $\mathbb{E}^N$. Let m be a point of M^n and U be a neighborhood of m. Let ξ be a unit normal vector field on U. Let

$$(e_1, e_2, \ldots, e_n)$$

be an orthonormal frame of TU and let

$$(\xi_1, \xi_2, \xi_3, \ldots, \xi_{N-n-1}, \xi_{N-n} = \xi)$$

be an orthonormal frame of $T^\perp U$. We can complete the previous frame to get, at each point (m, ξ), an orthonormal frame of $T_{(m,\xi)}\mathbb{E}^N \times \mathbb{S}^{N-1}$ adapted to M^n in the following way. The vectors

$$
\begin{aligned}
&\varepsilon_1 = \begin{pmatrix} e_1 \\ 0 \end{pmatrix}, \varepsilon_2 = \begin{pmatrix} e_2 \\ 0 \end{pmatrix}, \ldots, \varepsilon_n = \begin{pmatrix} e_n \\ 0 \end{pmatrix}, \\
&\varepsilon_{n+1} = \begin{pmatrix} \xi_1 \\ 0 \end{pmatrix}, \varepsilon_{n+2} = \begin{pmatrix} \xi_2 \\ 0 \end{pmatrix}, \ldots, \varepsilon_N = \begin{pmatrix} \xi_{N-n} \\ 0 \end{pmatrix}, \\
&\tilde{\varepsilon}_1 = \begin{pmatrix} 0 \\ e_1 \end{pmatrix}, \tilde{\varepsilon}_2 = \begin{pmatrix} 0 \\ e_2 \end{pmatrix}, \ldots, \tilde{\varepsilon}_n = \begin{pmatrix} 0 \\ e_n \end{pmatrix}, \\
&\tilde{\varepsilon}_{n+1} = \begin{pmatrix} 0 \\ \xi_1 \end{pmatrix}, \ldots, \tilde{\varepsilon}_{N-1} = \begin{pmatrix} 0 \\ \xi_{N-n-1} \end{pmatrix}
\end{aligned}
\tag{21.2}
$$

define an orthonormal frame of $T_{(m,\xi)}\mathbb{E}^N \times \mathbb{S}^{N-1}$. The $(N-1)$-form

$$\varepsilon_{M^n} = \begin{pmatrix} e_1 \\ D_{e_1}\xi \end{pmatrix} \wedge \begin{pmatrix} e_2 \\ D_{e_2}\xi \end{pmatrix} \wedge \ldots \wedge \begin{pmatrix} e_n \\ D_{e_n}\xi \end{pmatrix} \wedge \begin{pmatrix} 0 \\ \xi_1 \end{pmatrix} \wedge \begin{pmatrix} 0 \\ \xi_2 \end{pmatrix} \wedge \ldots \wedge \begin{pmatrix} 0 \\ \xi_{N-n-1} \end{pmatrix}$$

is a (global) volume form of $ST^\perp M^n$.

Let

$$(h^\xi_{ij}) = <D_{e_i}\xi, e_j>$$

be the second fundamental form in the direction ξ and let $\Xi_k(\xi)$ be the corresponding k^{th}-mean curvature (see (11.15)). Denoting by $<.,.>$ the scalar product on differential forms induced by the standard scalar product on $\mathbb{E}^N \times \mathbb{S}^{N-1}$, we have the following proposition.

Proposition 17. *At every point (m, ξ) of $ST^\perp M^n$,*

$$
\begin{cases}
<\mathcal{W}_k, \varepsilon_{M^n}>_{(m,\xi)} & = 0 \ if \ k < N-n-1, \\
<\mathcal{W}_{(N-n-1)}, \varepsilon_{M^n}>_{(m,\xi)} & = 1, \\
<\mathcal{W}_{(N-n-1+k)}, \varepsilon_{M^n}>_{(m,\xi)} & = \Xi_k(\xi).
\end{cases}
\tag{21.3}
$$

In particular,

$$< \mathcal{W}_{(N-1)}, \mathcal{E}_{M^n} >_{(m,\xi)} = \det h^{\xi}. \tag{21.4}$$

Consequently, integrating these forms on the unit normal bundle of the submanifold M^n, we obtain, via the Gauss equation, the classical invariants $\Phi_k(M)$ (we use the same notations as in Chap. 17):

$$< \mathbf{N}(M^n), \mathcal{W}_{N-n-1+k} >= \frac{n!}{(N-k)!k!s_{k-1}} \Phi_{N-n+k}(M^n). \tag{21.5}$$

We deduce the following theorem.

Theorem 68. *Let M^n be a smooth compact oriented submanifold of $\mathbb{E}^N$. Then*

$$\begin{cases} \mathcal{M}_{N-n-1+k}(M^n) & = \frac{n!}{(N-k)!s_{k-1}} \Phi_{N-n+k}(M^n), \forall k \geq 0, \\ \mathcal{M}_k(M^n) & = 0, \text{ in the other cases,} \end{cases} \tag{21.6}$$

where $\Phi_k(M^n)$ denotes the coefficients of the Steiner polynomial of M^n.

Remarks

1. In particular,

$$\mathcal{M}_{N-n-1}(M^n) = \frac{s_{N-n-1}}{s_{N-1}} \text{Vol}_n(M^n). \tag{21.7}$$

2. On the other hand, using the Gauss–Bonnet theorem and (21.6), we see that, when n is even,

$$\mathcal{M}_{N-1}(M^n) = \frac{b_N}{s_{n-1}} \chi(M^n). \tag{21.8}$$

3. If M^n has no boundary, we deduce that, if k is odd,

$$\mathcal{M}_k(M^n) = 0 \tag{21.9}$$

(this can be seen by using the skew-symmetry of $\mathcal{W}_k$: for every (m, ξ),

$$\mathcal{W}_{(N-n+2l)_{(m,-\xi)}} = -\mathcal{W}_{(N-n+2l)_{(m,\xi)}}). \tag{21.10}$$

4. In other words, we can simplify the previous result as follows.

Corollary 15. *Let M^n be a compact oriented n-dimensional submanifold of $\mathbb{E}^N$. Then, the Weyl tubes formula can be stated as follows:*

$$\text{Vol}_N(M^n_\varepsilon) = \sum_{k=0}^{k=N} \Phi_k(M^n)\varepsilon^k, \tag{21.11}$$

with

$$\Phi_k(M^n) = \mathbf{C}(k,n,N)\mathcal{M}_{k-1}(M^n), \tag{21.12}$$

where $\mathbf{C}(k,n,N)$ is a positive constant depending only on the dimensions.

21.1.2 The Case of Polyhedra

To evaluate the invariant forms on the normal cycle of a polyhedron in $\mathbb{E}^N$, we begin by evaluating them on the elementary simplices and then applying the fundamental formula (see (20.1) or (20.5)):

$$\mathbf{N}(\mathcal{A} \cup \mathcal{B}) = \mathbf{N}(\mathcal{A}) + \mathbf{N}(\mathcal{B}) - \mathbf{N}(\mathcal{A} \cap \mathcal{B}) \tag{21.13}$$

for every geometric subsets $\mathcal{A}, \mathcal{B}$ in $\mathbb{E}^N$. Consequently, we only need to evaluate the curvatures $\mathcal{M}_k$ on simplices.

Theorem 69. *Let σ^l be an l-simplex of $\mathbb{E}^N$. Then, for every k,*

$$\mathcal{M}_k(\sigma^l) = \sum_{\sigma^k \subset \sigma^l} \mathrm{Vol}_k(\sigma^k)(\sigma^k, \sigma^l)^*. \tag{21.14}$$

Proof of Theorem 69 Let F be an affine l-dimensional subspace of $\mathbb{E}^N$. Let $(e_1, e_2, \ldots, e_l)$ be an orthonormal frame of F and let ξ be a unit vector orthogonal to F. Complete $(e_1, e_2, \ldots, e_l, \xi)$ to obtain an orthonormal frame

$$(e_1, e_2, \ldots, e_l, \xi, \xi_2 \ldots, \xi_{N-l})$$

of $\mathbb{E}^N$.

Let m be a point of F. With the previous notation,

$$(\varepsilon) = \begin{pmatrix} e_1 \\ 0 \end{pmatrix}, \begin{pmatrix} e_2 \\ 0 \end{pmatrix}, \ldots, \begin{pmatrix} e_l \\ 0 \end{pmatrix}, \begin{pmatrix} 0 \\ \xi_2 \end{pmatrix}, \ldots, \begin{pmatrix} 0 \\ \xi_{N-l} \end{pmatrix}$$

is an orthonormal frame of $TN(F)$ at (m, ξ).

We have

$$\begin{cases} \mathcal{W}_k(\varepsilon) & = 0 \text{ if } k \neq N - l - 1, \\ \mathcal{W}_{N-l-1}(\varepsilon) & = 1. \end{cases} \tag{21.15}$$

Now, let Q be an open set of (the support of) $\mathbf{N}(F)$ of the form $U \times U'$, where U is an open set of F and U' is an open set of $F^\perp \cap \mathbb{S}^{N-1}$. We have

$$\begin{cases} < \mathcal{W}_k, Q > & = 0 \text{ if } k \neq N - l - 1, \\ < \mathcal{W}_{(N-l-1)}, Q > & = \mathrm{Vol}(U)\mathrm{Vol}(U'). \end{cases} \tag{21.16}$$

Note that the support of the normal cycle of an l-simplex can be decomposed into the disjoint union of open sets of the previous shape:

$$\mathrm{suppt}\, \mathbf{N}(\sigma^l) = \cup_{\sigma^k \subset \sigma^l} \mathrm{int}(\sigma^k) C_{\sigma^k}, \tag{21.17}$$

where C_{σ^k} is the cone which has been used for the definition of the exterior dihedral angle. Now, by Definition 51,

$$\mathcal{M}_k(\sigma^l) = <\mathbf{N}(\sigma^l), \mathcal{W}_k>.$$

The conclusion follows. $\square$

Remarks

1. Equation (21.14) must be compared with (16.10), which only deals with convex bodies.
2. If P is any n-dimensional polyhedron, we still have

$$\mathcal{M}_n(P) = <\mathbf{N}(P), \mathcal{W}_n> = \mathrm{Vol}_n(P). \tag{21.18}$$

3. On the other hand, an easy computation gives

$$\mathcal{M}_{N-1}(P) = <\mathbf{N}(P), \mathcal{W}_{N-1}> = \frac{N!}{n!} s_{N-1} \chi(P). \tag{21.19}$$

4. If P has no boundary, we have

$$<\mathbf{N}(P), \mathcal{W}_{(N-n+2l)}> = 0, \tag{21.20}$$

since $\mathcal{W}_{((N-n-1)+2l+1_{(m,-\xi)})} = -\mathcal{W}_{((N-n-1)+2l+1_{(m,\xi)})}$.

5. Of course, if P is the boundary of a convex domain, the results found here are consistent with Theorem 47:

$$<\mathbf{N}(P), \mathcal{W}_k> = \sum_{\sigma^{(N-j)} \subset \bar{\sigma}^k} (-1)^{k-(N-j)} \mathrm{Vol}(\sigma^{N-j}).(\sigma^{N-j}, \sigma^k)^*. \tag{21.21}$$

21.2 Continuity of the $\mathcal{M}_k$

Since we do not have a general construction of the normal cycle of *any* compact subset of $\mathbb{E}^N$, a general convergence theorem is up to now impossible. However, we can give a basic framework for such a result and apply it in special cases: if $(\mathcal{A}_q)_{q\in\mathbb{N}}$ is a sequence of *geometric* subsets admitting normal cycles $\mathbf{N}(\mathcal{A}_q)$, such that $(\mathbf{N}(\mathcal{A}_q))_{q\in\mathbb{N}}$ converges (as a sequence of currents for the weak topology) to a current $\mathbf{N}$ of $\mathbb{E}^N \times \mathbb{S}^{N-1}$, then by definition of the weak convergence,

$$\forall k \in (0,...,N), \lim_{q\to\infty} <\mathbf{N}(\mathcal{A}_q), \mathcal{W}_k> = <\mathbf{N}, \mathcal{W}_k>,$$

i.e.,

$$\forall k \in (0,...,N), \lim_{q\to\infty} \mathcal{M}_k(\mathcal{A}_q) = <\mathbf{N}, \mathcal{W}_k>.$$

But up to now, it is not known in full generality if $<\mathbf{N}, \mathcal{W}_k>$ can represent the kth-curvature of some *geometric* subset of $\mathbb{E}^N$.

21.3 Curvature Measures of Geometric Sets

Following Wintgen [83], Zähle [85, 86], and Fu [49], let us introduce the *curvature measures* in a very general setting, extending the framework of Federer [42] for subsets with positive reach, on which these measures were defined earlier in Chap. 18.

Definition 52. Let $\mathcal{A}$ be a (compact) geometric subset of $\mathbb{E}^N$. Then, the kth-curvature measure on $\mathbb{E}^N$ with respect to $\mathcal{A}$ is defined by

$$\mathcal{M}_k^{\mathcal{A}}(B) = < \mathbf{N}(\mathcal{A}), \chi(B \times \mathbb{S}^{N-1}) \mathcal{W}_k >$$

for any Borel subset $B \subset \mathbb{E}^N.^2$

To simplify the notations, we already used the letter $\mathcal{M}_k$ for defining the kth-curvature of a geometric subset. One has $\mathcal{M}_k(\mathcal{A}) = \mathcal{M}_k^{\mathcal{A}}(\mathcal{A})$ for every Borel subset $\mathcal{A}$ of $\mathbb{E}^N$.

Remark It is important to note that these measures are *signed* measures on $\mathbb{E}^N$, which coincide with the measures defined in Chap. 18 for subsets with positive reach.

21.4 Convergence and Approximation Theorems

Using the main results on convergence and approximations on normal cycles (see Theorems 66 and 67), we deduce immediately convergence and approximation results for the curvature measures associated to geometric sets. The following result has been proved by Fu [48].

Theorem 70. *Let M^n be a smooth (compact) submanifold of $\mathbb{E}^N$. Let $(P_p)_{p \in \mathbb{N}}$ be a sequence of (compact) polyhedra closely inscribed in M^n such that:*

1. *The Hausdorff limit of P_p is M^n when p tends to infinity.*
2. *The Hausdorff limit of ∂P_p is ∂M^n when p tends to infinity.*
3. *The fatness of P_p is uniformly bounded from below by a nonnegative constant:*

$$\exists c > 0 \text{ such that }, \forall p \geq 0, \Theta(P_p) > c > 0.$$

Then for each k, the sequence of curvature measures $\mathcal{M}_k^{P_p}$ weakly converges^3 to the curvature measure $\mathcal{M}_k^{M^n}$.

Proof of Theorem 70 The proof is (almost) obvious since, under our assumptions, Theorem 66 claims that the sequence of normal cycles $\mathbf{N}(P_p)$ tends to the normal cycle of M^n for the flat topology. For every Borel subset B of $\mathbb{E}^N$, one has

---

^{2} χ denotes here the characteristic function.

^{3} In the sense of Definition 17.

$$\mathcal{M}_k^{P_p}(B) = <\mathbf{N}(P_p), \chi(B \times \mathbb{E}^N)\mathcal{W}_k >$$

and

$$\mathcal{M}_k^{M^n}(B) = <\mathbf{N}(M^n), \chi(B \times \mathbb{E}^N)\mathcal{W}_k > .$$

By standard arguments,[4] we deduce that the sequence of curvature measures $\mathcal{M}_k^{P_p}$ weakly converges to the curvature measure $\mathcal{M}_k^{M^n}$.

We also get the following approximation result for geometric subsets. We use the notation of Sect. 20.4. We assume that $\mathcal{V}$ is a weakly regular geometric subset of $\mathbb{E}^N$, which bounds the geometric subset $\mathcal{K}$ of $\mathbb{E}^N$ in the sense of Definition 48. Moreover, we assume that $\mathcal{V}$ is closely near a smooth (closed) hypersurface W which bounds a compact domain K, as defined in Definition 33. Let B be a (connected) Borel subset of $\mathcal{V}$. $\square$

Theorem 71. *For all $k, 0 \leq k \leq N-1$,*

$$|\mathcal{M}_k^{\mathcal{K}}(B) - \mathcal{M}_k^K(\mathrm{pr}_W B)| \leq$$

$$(\delta_B + \alpha_B)[\frac{2(1 + |h_{\mathrm{pr}_W(B)}|)}{1 - \delta_B |h_{\mathrm{pr}_W(B)}|}]^{N-1} (\mathbf{M}(\mathbf{N}(\mathcal{K})_{|T_B \mathbb{E}^N}) + \mathbf{M}(\partial \mathbf{N}(\mathcal{K})_{|T_B \mathbb{E}^N})).$$

Proof of Theorem 71 By definition,

$$|\mathcal{M}_k^{\mathcal{K}}(B) - \mathcal{M}_k^K(B)| = <\mathbf{N}(\mathcal{K}) - \mathbf{N}(K), \chi(B \times \mathbb{S}^{N-1})\mathcal{W}_k > .$$

Now, by definition of the flat norm and from Theorem 67,

$$<\mathbf{N}(\mathcal{K}) - \mathbf{N}(K), \chi(B \times \mathbb{S}^{N-1})\mathcal{W}_k > \leq \mathcal{F}(N(\mathcal{K})_{|T_B \mathbb{E}^N} - N(K)_{|T_{\mathrm{pr}_W(B)} \mathbb{E}^N})$$

$$\leq (\delta_B + \alpha_B)[\frac{2(1 + |h_{\mathrm{pr}_W(B)}|)}{1 - \delta_B |h_{\mathrm{pr}_W(B)}|}]^{N-1} (\mathbf{M}(\mathbf{N}(\mathcal{K})_{|T_B \mathbb{E}^N}) + \mathbf{M}(\partial \mathbf{N}(\mathcal{K}_{|T_B \mathbb{E}^N}))). \quad \square$$

We end this section by remarking that Rataj and Zähle [68–70, 89] proved convergence results for curvature measures in a large frame, introducing in particular the *space of second-order rectifiable currents*. Moreover, instead of dealing with a sequence of polyhedra converging to a smooth submanifold, Brehm and Kühnel [22] obtained convergence results on curvatures for a sequence of smooth submanifolds converging to a polyhedron. See also [13] for the definition of Lipschitz–Killing invariants in a general setting.

---

[4] The only difficulty lies in the fact that the weak convergence of measures involves integration over *continuous* functions, while the flat convergence of currents involves C^∞-differential forms with compact support. The masses of the involved normal cycles being uniformly bounded by Proposition 15, one classically solves this difficulty by approximating continuous functions by a sequence of C^∞-functions. This remark is due to J. Fu in a private communication.

Chapter 22
Second Fundamental Measure

In Chap. 21, we have seen how to define generalized *Lipschitz–Killing curvature measures* on geometric subsets, which generalize the Lipschitz–Killing curvature measures on smooth surfaces. However, the geometry of a smooth submanifold needs more precise invariants, like principal directions, principal curvatures, and lines of curvatures, which are determined by the full knowledge of the second fundamental tensor. The goal of this chapter is to propose a definition of a second fundamental measure, which can be evaluated on nonsmooth objects using the theory of the normal cycle.

22.1 A Vector-Valued Invariant Form

In this section, we introduce an $(N-1)$-form on $\mathbb{E}^N \times \mathbb{S}^{N-1}$ depending (bi)linearly on two parallel vector fields defined on $\mathbb{E}^N$. In this context, it may be clearer to consider that $\mathbb{S}^{N-1}$ is the unit sphere of $\mathbb{E}^N$ and write $\mathbb{E}^N \times \mathbb{S}^{N-1} \subset \mathbb{E}^n \times \mathbb{E}^N \simeq T\mathbb{E}^N$. Moreover, we shall identify frequently the first and the second factor of $\mathbb{E}^N \times \mathbb{E}^N$. Let U be an open neighborhood of a point $(m, \xi) \in \mathbb{E}^n \times \mathbb{S}^{N-1}$. Once again, let $(e_1, ..., e_n)$ be a local orthonormal frame on $\mathbb{E}^N$ such that $e_N = \xi$. We use the notation of Chap. 21. For two fixed vectors $e_{i_0} \neq e_n$ and $e_{j_0} \neq e_n$, we define the $(n-1)$-form

$$\mathbf{h}_{i_0, j_0} = (-1)^{N-i_0} \varepsilon^1 \wedge ... \wedge \widehat{\varepsilon^{i_0}} \wedge ... \wedge \varepsilon^N \wedge \tilde{\varepsilon}_{j_0}.$$

When the two indices i_0 and j_0 vary, the previous formula defines a tensor of type $(0, 2)$, independent of the orthonormal local frame $(e_1, ..., e_N)$: if $X, Y \in \mathbb{E}^N, X = \sum_{i=1}^N X^i e_i, Y = \sum_{i=1}^N Y^i e_i$, then

$$\mathbf{h}(X, Y) = \sum_{i, j=1}^{N-1} X^i Y^j \mathbf{h}_{i, j}.$$

To simplify the notation, we put

$$\mathbf{h}(X,Y) = \mathbf{h}^{X,Y}.$$

Definition 53. The form $\mathbf{h}^{X,Y}$ is called the fundamental $(N-1)$-form associated to the couple (X,Y).

The tensor $\mathbf{h}$ has an interesting property involving the symplectic structure Ω on TM.

Proposition 18. *For every* X,Y *in* $\mathbb{E}^N$, *there exists an* $(N-3)$-*form* $\Psi^{X,Y}$ *on* $\mathcal{T}\mathbb{E}^n$ *such that*

$$\mathbf{h}^{X,Y} - \mathbf{h}^{Y,X} = \Omega \wedge \Psi^{X,Y}.$$

Proof of Proposition 18 Using the linearity, we shall deal with an orthonormal frame $(\varepsilon^i, \tilde{\varepsilon}^j)$. The only interesting case occurs when we consider two indices $\mathbf{h}_{i_0 j_0}$ with i_0 and j_0 different of n. In this case,

$$
\begin{aligned}
\mathbf{h}_{i_0 j_0} - \mathbf{h}_{j_0 i_0} &= (-1)^{N-i_0} \varepsilon^1 \wedge \ldots \wedge \widehat{\varepsilon^{i_0}} \wedge \ldots \wedge \varepsilon^j \wedge \ldots \wedge \hat{\varepsilon}^n \wedge \tilde{\varepsilon}_{j_0} \\
&\quad - (-1)^{N-j_0} \varepsilon^1 \wedge \ldots \wedge \hat{\varepsilon}^{j_0} \wedge \ldots \wedge \varepsilon^j \wedge \ldots \wedge \hat{\varepsilon}^n \wedge \tilde{\varepsilon}_{i_0} \\
&= \pm (\varepsilon^{i_0} \wedge \tilde{\varepsilon}_{i_0} + \varepsilon^{j_0} \wedge \tilde{\varepsilon}_{j_0}) \wedge (\Lambda_{k \neq i_0, j_0, n} \varepsilon^k) \\
&= \pm \Omega \wedge (\varepsilon^i \wedge \ldots \wedge \varepsilon^j). \quad \square
\end{aligned}
\tag{22.1}
$$

22.2 Second Fundamental Measure Associated to a Geometric Set

In the same flavor as Definition 51, we give the following result.

Definition 54. Let $\mathcal{K}$ be a compact geometric subset of $\mathbb{E}^N$ and let X,Y be any parallel vector fields lying in $\mathbb{E}^N$. Then, the second fundamental measure $\mathbf{h}_{\mathcal{K}}^{X,Y}$ associated to $\mathcal{K}$ in the directions X,Y is defined by

$$\mathbf{h}_{\mathcal{K}}^{X,Y}(B) = <\mathbf{N}(\mathcal{K}), \chi_{B \times \mathbb{E}^N} \mathbf{h}^{X,Y}>$$

for every Borel subset B of M.

Here is a remarkable symmetry property of $\mathbf{h}_{\mathcal{K}}$.

Proposition 19. *Let* $\mathcal{K}$ *be any geometric subset and let B be any Borel subset of M. Then,* $\mathbf{h}_{\mathcal{K}}^{X,Y}(B)$ *is symmetric in* X,Y.

Proof of Proposition 19 This is a direct consequence of the fact that normal cycles are Legendrian and then cancel the symplectic form Ω restricted to $\mathbb{E}^N \times \mathbb{S}^{N-1}$. Using Proposition 18 and its notations,

$$\mathbf{h}_{\mathcal{K}}^{X,Y}(B) - \mathbf{h}_{\mathcal{K}}^{Y,X}(B) = <\mathbf{N}(\mathcal{K}), \chi_{B \times \mathbb{E}^N} \Omega \wedge \Psi^{X,Y}> = 0. \quad \square$$

22.3 The Case of a Smooth Hypersurface

As we have seen, the goal of the second fundamental measure is to generalize the second fundamental form of a hypersurface. For further use, we shall extend a little the definition of the classical second fundamental form. Suppose that K is a compact domain of $\mathbb{E}^N$ whose boundary is an oriented (smooth) hypersurface of W. At each point p of W, we denote by $\tilde{h}_p$ the bilinear form

$$\tilde{h}_{M_p} : \mathbb{E}^N_p \times \mathbb{E}^N_p \to \mathbb{R},$$

defined by composition of the second fundamental form h_p of W at p with the projection pr_{T_pW} onto T_pW:

$$\tilde{h}_{W_p} = h_p \circ (\mathrm{pr}_{T_pW}, \mathrm{pr}_{T_pW}).$$

Here, $\tilde{h}_{W_p}$ is clearly null on the normal space of W at p. The only difference is that it is defined on the whole tangent space of $\mathbb{E}^N$ at p.

As in Chap. 21, we can give an explicit expression of $\mathbf{h}$ in particular cases. We need to introduce the Gauss map G associated to the immersion of W:

$$G : W \hookrightarrow \mathbb{E}^N \times \mathbb{E}^N,$$

defined by

$$G(m) = (m, \xi_m).$$

Proposition 20. *One has*

$$\forall X, Y \in TW, h(X, Y) dv_W = G^* \mathbf{h}(X, Y),$$

where dv_W denotes the volume form of W.

Proof of Proposition 20 Let $(e_1, ..., e_n)$ be a local frame of M such that $e_1, ..., e_{n-1}$ are tangent to W and e_n is normal to M. Let (e_{i_0}, e_{j_0}) be two vectors of this frame, different to e_n. Then

$$\begin{aligned}
G^* \mathbf{h}(e_{i_0}, e_{j_0})(e_1, ..., e_{n-1}) &= \mathbf{h}(e_{i_0}, e_{j_0})(dG(e_1), ..., dG(e_{n-1})) \\
&= \Theta_{j_0}(dG(e_{i_0})) \qquad\qquad (22.2) \\
&= h(e_{i_0}, e_{j_0}). \quad \square
\end{aligned}$$

An immediate corollary can be stated as follows. Let B be a Borel subset of W.

Corollary 16.

$$< ST^{\perp}_{B \cap W} W, \mathbf{h}^{\mathrm{pr}_{TW}X, \mathrm{pr}_{TW}Y} > = \int_{B \cap W} h(\mathrm{pr}_{TW}X, \mathrm{pr}_{TW}Y) dv_W. \qquad (22.3)$$

22.4 The Case of a Polyhedron

Next, we assume that P is an N-dimensional polyhedron of $\mathbb{E}^N$. We shall evaluate $\mathbf{h}_P^{X,Y}$ for any (parallel) vector fields $X, Y \in \mathbb{E}^N$. Since the normal cycle $N(P)$ can be decomposed as a sum of elementary currents, the support of which lies above each simplex of dimension $i, 1 \leq i \leq N - 1$, we shall evaluate $\mathbf{h}^{X,Y}$ above each simplex. If σ^k is any k-dimensional simplex of P, the support of $N(P)$ lying above σ^k is the product of σ^k by a portion of a vertical $(N - k - 1)$-sphere. In particular, the support of $N(P)_{|\sigma^{N-2}}$ is the product $\sigma^{N-2} \times C_\sigma$, where C is a portion of circle. Let $(e_1, ..., e_{N-2})$ be an orthonormal frame field tangent to σ^{N-2}. Any point of $\sigma^{N-2} \times C_\sigma$ is a couple (m, e_{N-1}), where m is a point of σ^{N-2} and e_{N-1} is a unit vector orthogonal to σ^{N-2}. With these notations, we have the following theorem.

Theorem 72. *For every Borel set* $B \subset \mathbb{E}^N$,

$$\mathbf{h}_P^{X,Y}(B) = \sum_{\sigma^{N-2} \subset \partial P} \int_{\sigma^{N-2} \cap B \times C} < X, e_{(N-1)} >< Y, e_{(N-1)} > .$$

Sketch of proof of Theorem 72 At each point of $ST\mathbb{E}^N$, the form $\mathbf{h}^{X,Y}$ is the wedge product of an $(N - 2)$-form tangent to the horizontal bundle and a 1-form tangent to the vertical bundle. Consequently, when we plug it in the normal cycle of P, the only non-null contribution is given by the $(N - 2)$-simplices of ∂P. The explicit computation gives trivially the theorem. $\square$

22.5 Convergence and Approximation

Using the same techniques as those developed in Chap. 20, we have a convergence result for the second fundamental measure of polyhedra and a general approximation result for the second fundamental measure of geometric sets.

Theorem 73. *Let* M^n *be a smooth submanifold of* $\mathbb{E}^N$. *Let* $(P_p)_{p \in \mathbb{N}}$ *be a sequence of polyhedra strongly inscribed in* M^n *such that:*

1. *The Hausdorff limit of* P_p *is* M^n *when* p *tends to infinity.*
2. *The Hausdorff limit of* ∂P_p *is* ∂M^n *when* p *tends to infinity.*
3. *The fatness of* P_p *is uniformly bounded by below by a nonnegative constant:*

$$\exists c > 0 \text{ such that } \forall p \geq 0, \Theta(P_p) > c > 0.$$

Then, the sequence of second fundamental measures $(\mathbf{h}_{P_p})_{p \in \mathbb{N}}$ *converges to the second fundamental measure* $\mathbf{h}_{M^n}$ *for the weak topology of measures.*

Corresponding to Theorem 71, we have the following re sult, using the notation of Chap. 21. We assume that $\mathcal{W}$ is a weakly regular geometric subset of $\mathbb{E}^N$ which bounds the geometric subset $\mathcal{K}$ of $\mathbb{E}^N$. Moreover, we assume that $\mathcal{W}$ is closely near a smooth (closed) hypersurface W which bounds a compact domain K.

Theorem 74. *For every Borel subset of* $\mathbb{E}^N$,

$$|\mathbf{h}_\mathcal{K}(B) - \mathbf{h}_K(B)| \leq$$

$$(\delta_B + \alpha_B)[\frac{2(1+|h_{\mathrm{pr}_W(B)}|)}{1-\delta_B|h_{\mathrm{pr}_W(B)}|}]^{N-1}(\mathbf{M}(\mathbf{N}(\mathcal{K})_{|T_B\mathbb{E}^N}) + \mathbf{M}(\partial\mathbf{N}(\mathcal{K}_{|T_B\mathbb{E}^N}))).$$

22.6 An Example of Application

We present here a concrete application in graphism. The goal is to visualize the local geometry of Michelangelo's David. Let us consider a given triangulation P of the head of David. Let B be a union of triangles around a vertex v^1 and compute the curvature tensor $\mathbf{h}_P(B)$. We get a 3×3 symmetric matrix which can be diagonalized. We deduce three orthogonal eigendirections. By analogy with the smooth case (where the normal direction to the surface is associated to a null eigenvalue), we define the *normal direction* of $B \cap P$ as the direction associated to the smallest eigenvalue. The two others are called the *principal directions*. They are associated to the two other eigenvalues, called the *principal curvatures* of $B \cap P$. We can now draw the directions of minimal curvatures and integrate them to obtain lines of curvatures (see Fig. 22.1).

Although the visual result seems fine, we must temper our optimism:

- If the curvature tensor has rank 1, then the estimated normal may lie anywhere in the plane orthogonal to the third eigendirection.
- The result depends obviously on the chosen union of triangles B.
- Finally, since we do not have any equation of a smooth surface underlying the triangulation,[2] our theorem of approximation (Theorem 74) does not give any information.

---

[1] More precisely, we take the 1-ring or the 2-ring around v.

[2] That is, a "smooth David."

Fig. 22.1 Here is an example of directions of minimal curvatures and lines of minimal curvatures, estimated on Michelangelo's David. The first image is the triangulation on which the second fundamental measure has been estimated. These images are courtesy of Digital Michelangelo Project, Stanford University, and courtesy of the I.N.R.I.A. project team Geometrica

Part VIII
Applications to Curves and Surfaces

Chapter 23
Curvature Measures in $\mathbb{E}^2$

23.1 Invariant Forms of $\mathbb{E}^2 \times \mathbb{S}^1$

The determination of the invariant 1-forms of $\mathbb{E}^2 \times \mathbb{S}^1$ can be easily deduced from Chap. 19. Let (m, ξ) be a point of $\mathbb{E}^2 \times \mathbb{S}^1$. Consider an orthonormal frame of $\mathbb{E}^2$ whose second vector is ξ: $(e_1, e_2 = \xi)$. Then, an orthonormal frame of $T_{(m,\xi)}\mathbb{E}^2 \times \mathbb{S}^1$ is given by the vectors

$$\varepsilon_1 = \begin{pmatrix} e_1 \\ 0 \end{pmatrix}, \quad \varepsilon_2 = \begin{pmatrix} e_2 = \xi \\ 0 \end{pmatrix}, \quad \tilde{\varepsilon}_1 = \begin{pmatrix} 0 \\ e_1 \end{pmatrix}.$$

We deduce from Theorem 63 that the space of invariant 1-forms on $\mathbb{E}^2 \times \mathbb{S}^1$ has dimension 2. By identifying as usual 1-forms with vectors, we conclude that, at a point $(p, \xi = e_2)$, the space of invariant 1-forms is spanned by the forms

$$\mathcal{W}_0 = \varepsilon_1, \quad \mathcal{W}_1 = \tilde{\varepsilon}_1.$$

23.2 Bounded Domains in $\mathbb{E}^2$

We now restrict our attention to (embedded) smooth curves or polygon lines with or without boundary embedded in $\mathbb{E}^2$. When they are closed, they can be considered as the boundary of a (bounded) domain. We study in detail their normal cycle.

23.2.1 The Normal Cycle of a Bounded Domain

Consider a domain $\mathcal{D}$ whose boundary is a closed (embedded) curve:

- If this curve is smooth, then the normal cycle $\mathbf{N}(\mathcal{D})$ is nothing but the cycle whose support is the unit normal bundle of $\mathcal{D}$. The orientation of $\mathbf{N}(\mathcal{D})$ is given by the

Fig. 23.1 The image by
the exponential map of the
support of the normal cycle of
a domain $\mathcal{D}$ of $\mathbb{E}^2$ is the curve
γ. The orientation is given by
transport of the orientation of
the boundary

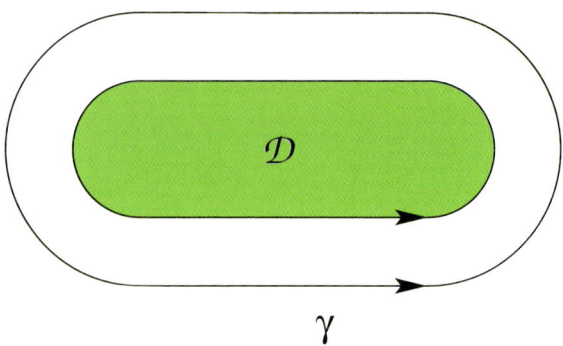

outward normal, via the exponential map. More precisely, if U is a "suitable"
neighborhood of $\mathcal{D}$, then the exponential map

$$\exp : TU \to U$$

given by

$$(x, \eta) \to x + \varepsilon \eta,$$

where ε is small enough, allows us to identify the support of $\mathbf{N}(\mathcal{D})$ with a closed
curve in $\mathbb{E}^2$. This closed curve has a canonical orientation given by its outward
normal, which induces an orientation of the support of $\mathbf{N}(\mathcal{D})$ (Fig. 23.1).

- Suppose now that $\mathcal{D}$ has a (closed) polygonal boundary P:

  - The simplest situation is the case where $\mathcal{D}$ is a (full) triangle. Then, $\mathbf{N}(\mathcal{D})$ can
    be decomposed as the sum of two currents, $\mathbf{N}^e(\mathcal{D})$ over the edges and $\mathbf{N}^v(\mathcal{D})$
    over the vertices as follows:

    1. Over an edge e, the support of the normal cycle is

    $$e \times \xi_e \in \mathbb{E}^2 \times \mathbb{S}^1,$$

      where ξ_e denotes the outward unit normal to e.
    2. Over a vertex v, the support of the normal cycle is

    $$v \times C_v \in \mathbb{E}^2 \times \mathbb{S}^1,$$

      where C_v denotes the arc of the unit circle joining the outward normals of
      the edges incident at v.

    The orientation of $\mathbf{N}(\mathcal{D})$ is determined as before (Fig. 23.2).

  - Let us generalize this situation: suppose that the boundary of $\mathcal{D}$ is a closed
    (embedded) polygon line. One can recover the normal cycle of $\mathcal{D}$ by decom-
    posing $\mathcal{D}$ into triangles and apply the additivity formula (20.1). It is easy to
    check that $\mathbf{N}(\mathcal{D})$ can be decomposed as the sum of two currents, $\mathbf{N}^e(\mathcal{D})$ over
    the edges and $\mathbf{N}^v(\mathcal{D})$ over the vertices as follows:

Fig. 23.2 The image by the exponential map of the support of the normal cycle of a full triangle $\mathcal{D}$ of $\mathbb{E}^2$ is the curve γ. The orientation is given by transport of the orientation of the boundary

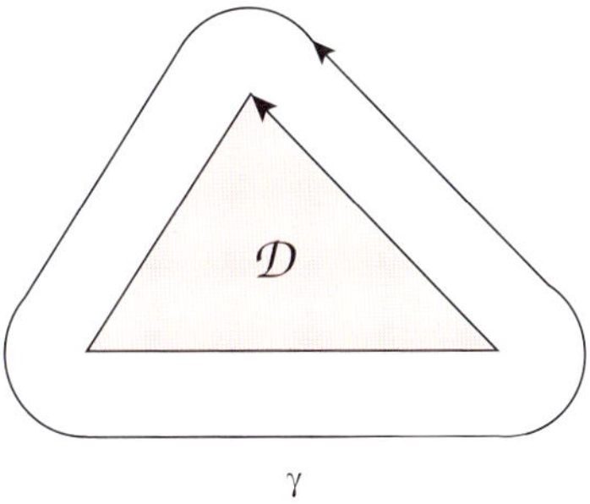

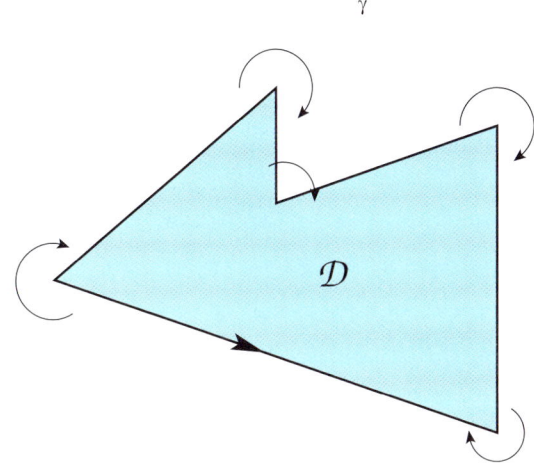

Fig. 23.3 The image by the exponential map of the support of the normal cycle "over the vertices" of a polygonal domain $\mathcal{D}$ of $\mathbb{E}^2$, with the correct orientation (for clarity in the picture, we do not draw the support of the normal cycle of $\mathcal{D}$ over the edges)

1. Over an edge e of the boundary of $\mathcal{D}$, the support of $\mathbf{N}^e(\mathcal{D})$ is the segment $e \times \xi_e$, where ξ_e is the outward normal to e.
2. Over a vertex v of the boundary of $\mathcal{D}$, the support of $\mathbf{N}^v(\mathcal{D})$ is the segment $v \times C_v$, where C_v denotes the arc of the unit joining the outward normals of the edges of the boundary of $\mathcal{D}$ incident at v.
3. All other contributions of the vertices and edges in the interior of $\mathcal{D}$ cancel (Fig. 23.3).

23.2.2 The Mass of the Normal Cycle of a Domain in $\mathbb{E}^2$

Using Sect. 23.2.1, it is easy to evaluate the mass of the normal cycle of a domain $\mathcal{D}$ in $\mathbb{E}^2$. It will be interesting to localize our formulas, to allow us to apply the approximation Theorem 71:

- If the boundary of $\mathcal{D}$ is a smooth (closed) curve γ, then the mass $\mathbf{M}(\mathbf{N}(\mathcal{D}))$ is the length of its normal bundle:

$$\mathbf{M}(\mathbf{N}(\mathcal{D})) = \int_\gamma (1 + 2k + k^2)\,ds,$$

where k denotes the curvature of γ.

Let us localize this formula. Although we could take any Borel subset, we restrict our attention to Borel subsets B of γ:

$$\mathbf{M}(\mathbf{N}(\mathcal{D})_{B \times \mathbb{E}^3}) = \int_{\gamma \cap B} (1 + 2k + k^2) ds.$$

- If the boundary of $\mathcal{D}$ is a (closed) polygonal line P, then the mass $\mathbf{M}(\mathbf{N}(\mathcal{D}))$ can be computed by decomposing $\mathbf{N}(\mathcal{D})$. Clearly, $\mathbf{M}(\mathbf{N}(\mathcal{D}))$ is the sum of the length of the edges of P and the absolute value of the angles $|\angle(v)|$ at each vertex v between the normals of two consecutive edges:

$$\mathbf{M}(\mathbf{N}(\mathcal{D})) = l(P) + \sum_{v \in P} |\angle(v)|.$$

Of course, if we restrict our computation to B, then

$$\mathbf{M}(\mathbf{N}(\mathcal{D})_{|B \times \mathbb{E}^2}) = l(P) + \sum_{v \in P \cap B} |\angle(v)|.$$

23.3 Plane Curves

23.3.1 The Normal Cycle of an (Embedded) Curve in $\mathbb{E}^2$

- If the curve is smooth, its normal cycle is nothing but its unit normal bundle (Fig. 23.4).
- Let us now consider a planar polygon. To compute its normal cycle, we need to compute the normal cycles of a point and a segment and then use the additivity formula (20.1):

  – If p is reduced to a point of $\mathbb{E}^2$, $\mathbf{N}(p)$ is the current whose support is the (oriented) curve $\{p\} \times \mathbb{S}^1$.

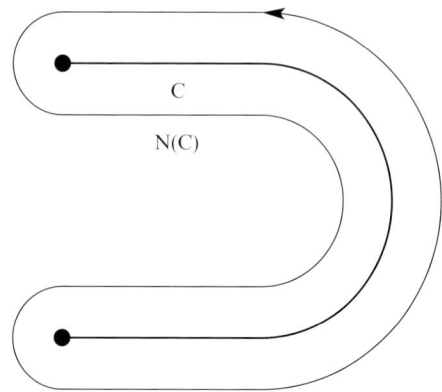

Fig. 23.4 The image by the exponential map of the support of the normal cycle $N(C)$ of a smooth curve C of $\mathbb{E}^2$. Note that $N(C)$ has a canonical orientation, which does not depend on the orientation of C

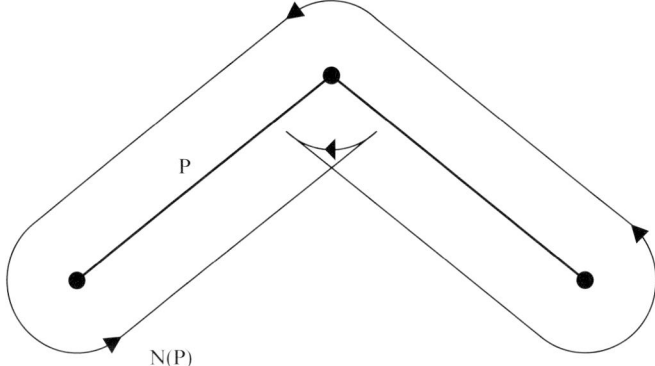

Fig. 23.5 The image by the exponential map of the support of the normal cycle $N(P)$ of the union of two segments P in $\mathbb{E}^2$

- If $I = [\alpha, \beta]$ is a segment of $\mathbb{E}^2$, $\mathbf{N}(I)$ is the current whose support is the (oriented) curve composed of two half-circles $\alpha \times C_\alpha$ and $\beta \times C_\beta$ and two segments $I \times \xi$ and $I \times (-\xi)$, where ξ is a unit vector normal to a.

As before, the orientation is naturally given via the exponential map. Consequently, by using the additivity formula (20.1), one can compute the normal cycle of any (embedded) polygon in $\mathbb{E}^2$ (Fig. 23.5):

- Over a vertex v which is not the boundary of the polygon, the support of the normal cycle is the (oriented) curve composed of two arcs of a circle of length $\angle(v)$, where $\angle(v)$ is the angle of the normals of the two edges with common vertex v.
- Over a vertex which is on the boundary of the polygon, the support of the normal cycle is an (oriented) half-circle.
- Over an edge e, the support of the normal cycle is reduced to two (oriented) segments whose length is $l(e)$.

The orientation is given as before via the exponential map (Fig. 23.6).

23.3.2 The Mass of the Normal Cycle of a Curve in $\mathbb{E}^2$

- If the curve C is smooth (regular and closed), then the mass of its normal cycle (i.e., the mass of its normal bundle) is its length. Then:

  - If C is closed,

$$\mathbf{M}(\mathbf{N}(C)) = 2 \int_C (1 + 2k + k^2)ds. \tag{23.1}$$

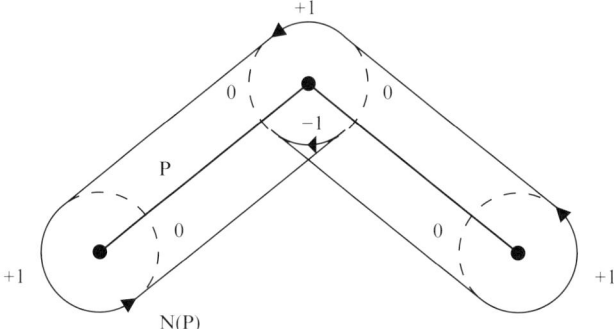

Fig. 23.6 The explicit computation of the normal cycle $\mathbf{N}(P)$ of a union P of two segments of $\mathbb{E}^2$. The integration of any 1-differential form over $\mathbf{N}(P)$ must be taken over the oriented plain closed curve, taking care of the orientation, as shown by the picture

- If C is not closed,

$$\mathbf{M}(\mathbf{N}(C)) = 2\int_C (1 + 2k + k^2)ds + 2\pi,$$

   since one must add to (23.1) the mass of the normal cycle "over the end points," i.e., the length of two half-circles.

- If the curve is a polygonal line P, then the mass of its normal cycle is the sum of the mass of the normal cycle "over the edges" and the mass of the normal cycle "over the vertices":

   - If P is closed,

$$\mathbf{M}(\mathbf{N}(P)) = 2l(P) + \sum_{v \text{ vertices of } P} \angle(v). \qquad (23.2)$$

   - If P is not closed,

$$\mathbf{M}(\mathbf{N}(P)) = 2l(P) + \sum_{v \text{ internal vertices of } P} \angle(v) + 2\pi. \qquad (23.3)$$

23.4 The Length of Plane Curves

We now plug the invariant 1-forms $\mathcal{W}_0, \mathcal{W}_1$ onto the normal cycles of curves.

23.4.1 Smooth Curves

Consider a (regular) smooth curve $\gamma: [a,b] \to \mathbb{E}^2$ embedded in $\mathbb{E}^2$. One can recover its length by integrating $\mathcal{W}_0$ over its unit normal bundle $N(\gamma)$. In fact, $N(\gamma)$ can be decomposed into three parts: $\gamma(a) \times C_1$, $\gamma(b) \times C_2$, over the boundaries a and b, where C_1 and C_2 are two half-circles, and two curves $\gamma(u) \times \xi(u)$ and $\gamma(u) \times (-\xi)(u)$, where ξ is the unit normal vector field over γ. Then

$$< \mathbf{N}(\gamma), \mathcal{W}_0 > = < C_1, \mathcal{W}_0 > + < C_2, \mathcal{W}_0 > + 2 \int_a^b |\gamma'(u)| du.$$

The expression for $\mathcal{W}_0$ shows that the two first terms on the right-hand side cancel. The third term is exactly twice the length of γ. Finally,

$$\mathcal{M}_0(\gamma) = < \mathbf{N}(\gamma), \mathcal{W}_0 > = 2l(\gamma).$$

23.4.2 Polygon Lines

- Consider now a segment I of $\mathbb{E}^2$. It is clear that

$$\mathcal{M}_0(I) = < \mathbf{N}(I), \mathcal{W}_0 > = 2l(I).$$

- If a polygon is the union of two segments I_1 and I_2, then

$$\mathbf{N}(I_1 \cup I_2) = \mathbf{N}(I_1) + \mathbf{N}(I_2) - \mathbf{N}(I_1 \cap I_2).$$

Here, $I_1 \cap I_2$ is reduced to a point v and the support of the normal cycle of v is $v \times \mathbb{S}^1$, on which $\mathcal{W}_0$ is null. This implies that

$$\mathcal{M}_0(I_1 \cup I_2) = < \mathbf{N}(I_1 \cup I_2), \mathcal{W}_0 > = l(I_1) + l(I_2) + 0 = l(I_1 \cup I_2).$$

- More generally, if P is any polygon embedded in $\mathbb{E}^2$,

$$\mathcal{M}_0(P) = < \mathbf{N}(P), \mathcal{W}_0 > = 2l(P).$$

23.5 The Curvature of Plane Curves

23.5.1 Smooth Curves

Consider a regular smooth curve $\gamma : [0, l] \to \mathbb{E}^2$ (parametrized by the arc length) embedded in $\mathbb{E}^2$. We shall evaluate $< \mathbf{N}(\gamma), \mathcal{W}_1 >$:

- If the curve is closed, then its normal cycle is the sum of two cycles: $\mathbf{N}_1$ parametrized by $(\gamma(s), \xi(s))$ and $\mathbf{N}_2$ parametrized by $(\gamma(s), -\xi(s))$, where $s \in [0, l]$, ξ being the unit normal vector field of γ. Now,

$$(\gamma, \xi)'(s) = (t, \xi') = (t, -kt),$$

from which we deduce (taking suitable orientations) that

$$< \mathbf{N}_1, \mathcal{W}_1 >= - \int_0^l k(s)ds.$$

A similar computation shows that

$$< \mathbf{N}_2, \mathcal{W}_1 >= \int_0^l k(s)ds,$$

from which we deduce that

$$< \mathbf{N}(\gamma), \mathcal{W}_1 >=< \mathbf{N}_1 + \mathbf{N}_2, \mathcal{W}_1 >= 0.$$

This means that to recover the integral of the curvature over γ, one needs to consider only one component of the support of the normal cycle of γ. In other words, suppose that γ bounds a domain $\mathcal{D}$. Then, the support of its normal cycle consists only of one curve (for instance $\mathbf{N}_1$). Then

$$< \mathbf{N}_1, \mathcal{W}_1 >=< \mathbf{N}(\mathcal{D}), \mathcal{W}_1 >= \int_{\partial \mathcal{D}} k(s)ds = \int_\gamma k(s)ds.$$

Consequently,

$$\mathcal{M}_1(\mathcal{D}) =< \mathbf{N}(\mathcal{D}), \mathcal{W}_1 >= \int_\gamma k(s)ds.$$

- Suppose now that γ is not closed, i.e., $\gamma(0) = \alpha, \gamma(l) = \beta, \alpha \neq \beta$. Then, the support of its normal cycle is a closed curve, composed of two half-circles C_α and C_β, one over each end point, and two curves $\mathbf{N}_1$ and $\mathbf{N}_2$ as before. Then, the same argument as previously shows that

$$< \mathbf{N}_1 + \mathbf{N}_2, \mathcal{W}_1 >= \pi + \pi = 2\pi.$$

Once again, we do not recover directly the curvature of the curve. We let the reader find a way to "cut correctly the normal cycle" and find a good result.

23.5.2 Polygon Lines

Consider a polygon line P embedded in $\mathbb{E}^2$. Let us evaluate

$$< \mathbf{N}(P), \mathcal{W}_1 > .$$

- Suppose that P is closed. The support of $\mathbf{N}(P)$ has two connected components. A direct computation shows that, over a vertex v, the support of the normal cycle $\mathbf{N}^v$ is composed of two arcs of circles of the same length $\angle(v)$, where $\angle(v)$ is the angle at v enclosed by the normal vectors of the two edges with common vertex v. Since their orientation is opposite,

$$< \mathbf{N}^v(P), \mathcal{W}_1 >= 0$$

for every vertex v. Moreover, if e is an edge, the support of the normal cycle $\mathbf{N}^e(P)$ restricted to e is the union of two segments, on which the integral of $\mathcal{W}_1$ is trivially null. Then

$$< \mathbf{N}^e(P), \mathcal{W}_1 >= 0,$$

from which we deduce immediately that

$$< \mathbf{N}(P), \mathcal{W}_1 >= 0.$$

The situation is similar to that of closed smooth curves: the integral of the form $\mathcal{W}_1$ over the normal cycle of P does not give any information on its geometry. However, consider the domain $\mathcal{D}$ bounded by P. One has

$$< \mathbf{N}(\mathcal{D}), \mathcal{W}_1 >= \sum_{v \text{ vertex } of P} \angle(v).$$

• Suppose now that P is not closed. Over each end point, the support of $\mathbf{N}^v(P)$ is a half-circle. An argument similar to the previous one gives immediately

$$< \mathbf{N}(P), \mathcal{W}_1 >= 2\pi.$$

The reader can now recover the sum of the angles at the vertices by integrating $\mathcal{W}_1$ over a "part" of the normal cycle $\mathbf{N}(P)$. If one defines the current $\mathbf{N}^+(P)$ as the "part of the normal cycle whose support is on one side of P," then one finds that

$$< \mathbf{N}^+(P), \mathcal{W}_1 >= \sum_{v \text{ vertex of } P} \angle(v) + \pi.$$

Consequently, it is natural to define the curvature $\mathcal{M}_1(P)$ of a planar polygon P by

$$\mathcal{M}_1^+(P) = \sum_{v \text{ vertex of } P} \angle(v).$$

Chapter 24
Curvature Measures in $\mathbb{E}^3$

24.1 Invariant Forms of $\mathbb{E}^3 \times \mathbb{S}^2$

Using Chap. 19, we can describe the space of invariant 2-forms of $\mathbb{E}^3 \times \mathbb{S}^2$. Let (m, ξ) be a point of $\mathbb{E}^3 \times \mathbb{S}^2$. Consider an orthonormal frame of $\mathbb{E}^3$ whose third vector is ξ:

$$(e_1, e_2, e_3 = \xi).$$

An orthonormal frame of $T_{(m,\xi)} \mathbb{E}^3 \times \mathbb{S}^2$ is given by the vectors

$$\varepsilon_1 = \begin{pmatrix} e_1 \\ 0 \end{pmatrix}, \varepsilon_2 = \begin{pmatrix} e_2 \\ 0 \end{pmatrix}, \varepsilon_3 = \begin{pmatrix} e_3 = \xi \\ 0 \end{pmatrix}, \widetilde{\varepsilon}_1 = \begin{pmatrix} 0 \\ e_1 \end{pmatrix}, \widetilde{\varepsilon}_2 = \begin{pmatrix} 0 \\ e_2 \end{pmatrix}. \qquad (24.1)$$

Still using Theorem 63 and the identification of 1-forms with vectors, we see that the space of invariant 2-forms of $\mathbb{E}^3 \times \mathbb{S}^2$ has dimension 4. At a point

$$(p, \xi = e_3),$$

it is spanned by the four 2-forms

$$\begin{aligned} \mathcal{W}_0 &= \varepsilon_1 \wedge \varepsilon_2, \quad \mathcal{W}_1 = \varepsilon_1 \wedge \widetilde{\varepsilon}_2 + \widetilde{\varepsilon}_1 \wedge \varepsilon_2, \\ \mathcal{W}_2 &= \widetilde{\varepsilon}_1 \wedge \widetilde{\varepsilon}_2, \quad \mathcal{W}_3 = \Omega_{|\mathbb{E}^3 \times \mathbb{S}^2} = \varepsilon_1 \wedge \widetilde{\varepsilon}_1 + \varepsilon_2 \wedge \widetilde{\varepsilon}_2. \end{aligned} \qquad (24.2)$$

24.2 Space Curves and Polygons

24.2.1 The Normal Cycle of Space Curves

Once again, the normal cycle of a smooth curve is nothing but its unit normal bundle. Let us now consider a polygon in $\mathbb{E}^3$. To compute its normal cycle, we compute the normal cycles of a point and a segment and then use the additivity formula:

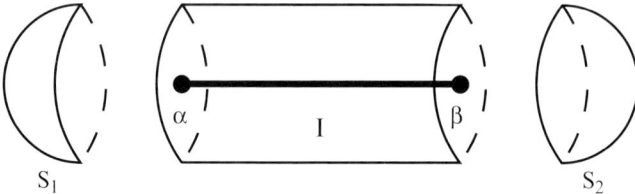

Fig. 24.1 The support of the normal cycle of the segment I can be decomposed into two half-spheres S_1 and S_2 and one cylinder $I \times \mathbb{S}^1$

- If p is reduced to a point of $\mathbb{E}^3$, $\mathbf{N}(p)$ is the current whose support is the (oriented) surface $p \times \mathbb{S}^2$.
- If $I = [\alpha, \beta]$ is a segment of $\mathbb{E}^3$, $\mathbf{N}(I)$ is the current whose support is the union of three (oriented) surfaces: $\alpha \times S_1$, $\beta \times S_2$, where S_1 and S_2 are two half-spheres (of dimension 2 and radius 1), and the cylinder $I \times \mathbb{S}^1$ (Fig. 24.1).

As above, the supports of normal cycles can be identified with surfaces of $\mathbb{E}^3$ via the exponential map.

Now, using the additivity formula (20.1), the normal cycle of any polygon in $\mathbb{E}^3$ can be computed:

- Over a vertex v which is not on the boundary of the polygon, the support of the normal cycle is composed of two (oriented) portions of spheres, delimited by the two 2-planes orthogonal to the edges whose common vertex is v.
- Over a vertex which is on the boundary of the polygon, the support of the normal cycle is an (oriented) half-sphere.
- Over (the interior of) an edge, the support of the normal cycle is an (oriented) cylinder.

In Sect. 24.2.2, we evaluate the previous invariant forms on the normal cycles of curves γ and polygons P. Note that

$$< \mathbf{N}(\gamma), \mathcal{W}_0 >=< \mathbf{N}(P), \mathcal{W}_0 >= 0,$$

which is compatible with the general theory.

24.2.2 The Length of Space Curves

We can recover the length of a smooth curve or a polygon by using the normal cycles. Consider a smooth curve γ and a polygon P of $\mathbb{E}^3$. A direct computation shows that

$$< \mathbf{N}(\gamma), \mathcal{W}_1 >= 2\pi l(\gamma),$$
$$< \mathbf{N}(P), \mathcal{W}_1 >= 2\pi l(P).$$

24.2.3 The Curvature of Space Curves

Consider a smooth (regular) curve γ and a polygon P of $\mathbb{E}^3$. A direct computation shows that:

- If γ is closed,
$$< \mathbf{N}(\gamma), \mathcal{W}_2 >= 0.$$

  If P is closed,
$$< \mathbf{N}(P), \mathcal{W}_2 >= 0.$$

- If γ is not closed,
$$< \mathbf{N}(\gamma), \mathcal{W}_2 >= 4\pi.$$

  If P is not closed,
$$< \mathbf{N}(P), \mathcal{W}_2 >= 4\pi.$$

Consequently, the curvature of γ or P cannot be recovered by a direct integration over the normal cycle of γ and P.

Suppose that the curve γ is closed and bounds a smooth (oriented) domain D. Then, a direct (and classical) computation gives

$$< \mathbf{N}(D), \mathcal{W}_2 >= 2 \int_D G da + 2 \int_\gamma \kappa^g(s) ds = 4\pi, \qquad (24.3)$$

where G is the Gauss curvature of D and κ^g is the geodesic curvature[1] of γ considered as a curve on D. The last equality of (24.3) comes from the Gauss–Bonnet theorem (see (3.15) and Theorem 22) and the fact that the Euler characteristic of a domain is 1.

Consider now a closed polygon P. Suppose that P bounds an (oriented) triangulated domain D. Then, P has a canonical orientation, and a direct computation shows that

$$< N(D), \mathcal{W}_2 >= 2 \sum_{v \in \text{ vertex of int}(D)} (2\pi - \alpha_v)$$
$$+ 2 \sum_{v \in \text{ vertex of } P} \beta_v \qquad (24.4)$$
$$= 4\pi,$$

where α_v denotes the solid angle at the interior vertex v and β_v denotes the *external* angle of the (oriented) consecutive edges e_{1_v}, e_{2_v} of P incident to the vertex $v \in P$. The sign of β_v is the sign of the determinant of the three vectors (e_{1_v}, e_{2_v}, ξ), where ξ is the oriented normal to the domain D (Fig. 24.2).

Again, the last equality comes from the fact that the Euler characteristic of a domain is 1.

---

[1] The geodesic curvature κ^g of the curve γ (parametrized by the arc length) defining the boundary of the domain D is defined as follows: at each point $\gamma(s)$, let $n(s)$ be the tangent vector orthogonal to $\gamma'(s)$ such that $\gamma'(s), n(s) >$ is a direct frame. Then, $\kappa^g(s) =< \gamma''(s), n >_s$ (see [40, 76] for instance).

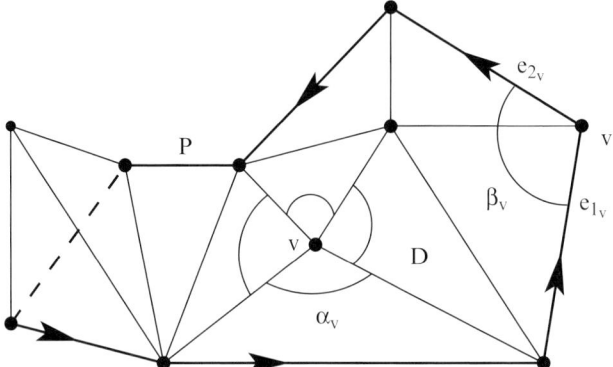

Fig. 24.2 At an interior vertex v, α_v is the solid angle of v. At a vertex v belonging to the boundary P of the domain, β_v is the external angle between the two incident edges belonging to P

Formulas (24.3) and (24.4) show that the curvature of a closed polygon can be defined as *the sum of the angles done by two consecutive edges*, by analogy with the smooth definition. Note, however, that we do not mimic the curvature of a smooth curve in $\mathbb{E}^3$, but the *geodesic curvature* of a smooth curve on a surface.

The reader will localize these formulas without difficulties.

24.3 Surfaces and Bounded Domains in $\mathbb{E}^3$

We deal here with smooth surfaces or two-dimensional polyhedra with or without boundaries, embedded in $\mathbb{E}^3$. When they are closed, they can be considered as boundaries of (bounded) domains. By a technique similar to that used for curves, we study their normal cycles.

24.3.1 The Normal Cycle of a Bounded Domain

Let $\mathcal{C}$ be a domain of $\mathbb{E}^3$ whose boundary is a closed surface:

- If the surface is smooth, then its normal cycle $\mathbf{N}(\mathcal{C})$ is nothing but the cycle whose support is the unit normal bundle of $\mathcal{C}$.
- We suppose now that the boundary of $\mathcal{C}$ is a (closed) two-dimensional (triangulated) polyhedron T:

  – Note that if $\mathcal{C}$ is a simplex, then it is convex and its normal cycle is the cycle of $\mathbb{E}^3 \times \mathbb{S}^2$ defined by the (smooth oriented closed) surface of $\mathbb{E}^3 \times \mathbb{S}^2$ given by

$$\{(m, \xi) \in \mathbb{E}^3 \times \mathbb{S}^2 \text{ such that } <\overrightarrow{mn}, \xi > \le 0, \forall n \in \mathcal{C}\}.$$

  – In general, $\mathcal{C}$ is a union of simplices. Hence, its normal cycle can be obtained by computing the normal cycle of each simplex and by using the additivity formula (20.1). Consequently, $\mathbf{N}(\mathcal{C})$ can be decomposed as a sum of three currents: $\mathbf{N}^f(\mathcal{C})$ over the faces, $\mathbf{N}^e(\mathcal{C})$ over the edges, and $\mathbf{N}^v(\mathcal{C})$ over the vertices of $\mathcal{C}$:

1. Over a face f, the support of $\mathbf{N}^f(\mathcal{C})$ is the set of points

$$\{(m,\xi) \in \mathbb{E}^3 \times \mathbb{S}^2, m \in f, \xi \text{ unit vector normal to } \mathcal{C} \text{ at } m.\}$$

2. Over an edge e, the support of $\mathbf{N}^e(\mathcal{C})$ is the portion

$$e \times C \in \mathbb{E}^3 \times \mathbb{S}^2$$

of a cylinder, where C is the arc of a circle of $\mathbb{S}^2$ defined by the unit vectors delimited by the normals of the two faces incident to e. If e is convex with respect to $\mathcal{C}$, then $\mathbf{N}^e(\mathcal{C})$ above e can be identified with the current defined by integration over this portion of the cylinder, endowed with the canonical orientation and with multiplicity $+1$.

3. Over a vertex v, the support of $\mathbf{N}^v(\mathcal{C})$ is a portion of the unit sphere $v \times \mathbb{S}^2$, and $\mathbf{N}^v(\mathcal{C})$ above v can be identified with a linear combination of currents defined by integration over portions of the unit sphere over v, endowed with correct multiplicity. More precisely, the multiplicity of each point h of the unit sphere over v is the index of v with respect to h, i.e., the integer

$$\mu(v,h) = \chi(St^+(v,h)),$$

where $\chi(St^+(v,h))$ is the Euler characteristic of the union of relative interiors of cells of $\mathcal{C}$ incident to v and lying in the half-plane $\{x :< \vec{vx}, h > \geq 0\}$ (see [8] for a general study).

24.3.2 The Mass of the Normal Cycle of a Domain in $\mathbb{E}^3$

Using Sect. 24.3.1, we can easily evaluate the mass of the normal cycle of $\mathcal{C}$. To apply Theorem 70, we localize our computations:

- If the boundary of $\mathcal{C}$ is a smooth surface S, then the mass $\mathbf{M}(\mathbf{N}(\mathcal{C}))$ is the area of its (two-dimensional) unit normal bundle. It is interesting to localize this result as follows: let B be a Borel subset of S. Then, $\mathbf{M}(\mathbf{N}(\mathcal{C})_{|B \times \mathbb{E}^3})$ is the area of the restriction (to $B \times \mathbb{E}^3$) of the support of $\mathbf{N}(\mathcal{C})$.

- If the boundary of $\mathcal{C}$ is a triangulation T, then the mass $\mathbf{M}(\mathbf{N}(\mathcal{C}))$ can be computed by decomposing $\mathbf{N}(\mathcal{C})$. Let us localize the computation by considering the interior B of a finite union of triangles of T, and evaluating the mass of $\mathbf{N}(\mathcal{C})_{|B \times \mathbb{E}^3}$. Using the previous decomposition of the normal cycle, we decompose this mass into three terms:

1. *The mass over the faces*: $\mathbf{M}(\mathbf{N}^f(\mathcal{C})_{|B \times \mathbb{E}^3})$ is the area $\mathcal{A}(B)$ of B.
2. *The mass over the edges*: $\mathbf{M}(\mathbf{N}^e(\mathcal{C})_{|B \times \mathbb{E}^3})$ is the sum over all the edges e of B of the products $l(e)|\angle(e)|$, where $l(e)$ denotes the length of the edge e and $|\angle(e)|$ denotes the absolute value of the angle between the normals of two incident faces to e lying in B.
3. *The mass over the vertices*: we begin with the evaluation of the mass $\mathbf{M}(\mathbf{N}^v(\mathcal{C})_{|B \times \mathbb{E}^3})$ of the normal cycle over a vertex $v \in B$. Decomposing the sphere $v \times \mathbb{S}^2$ over v as a disjoint union of Borel subsets U_{v_i} on which the multiplicity $\mu(v, h)$ is constant when h varies, and denoting by μ_{v_i} this constant, we have

$$\mathbf{M}(\mathbf{N}^v(\mathcal{C})_{|v \times \mathbb{E}^3}) = \sum_i |\mu_{v_i}| \mathcal{A}(U_{v_i}) \tag{24.5}$$

and

$$\mathbf{M}(\mathbf{N}^v(\mathcal{C})_{|B \times \mathbb{E}^3}) = \sum_{v \in B} \mathbf{M}(\mathbf{N}^v(\mathcal{C})_{|v \times \mathbb{E}^3}). \tag{24.6}$$

Finally, with the previous notations,

$$\mathbf{M}(\mathbf{N}(\mathcal{C})_{|B \times \mathbb{E}^3}) = \mathcal{A}(B) + \sum_{e \subset B} l(e)|\angle(e)| + \sum_{v \in B} \sum_i |\mu_{v_i}| \mathcal{A}(U_{v_i}). \tag{24.7}$$

Globally, by taking $B = T$, we obtain

$$\mathbf{M}(N(\mathcal{C})) = \mathcal{A}(T) + \sum_e l(e)|\angle(e)| + \sum_v \sum_i |\mu_{v_i}| \mathcal{A}(U_{v_i}). \tag{24.8}$$

24.3.3 The Curvature Measures of a Domain

Let $\mathcal{C}$ be a compact domain of $\mathbb{E}^3$ and let B be a (bounded) Borel subset of $\mathbb{E}^3$. To simplify, we assume that $B \subset \partial \mathcal{C}$. Let us evaluate the invariant forms $\mathcal{W}_0, \mathcal{W}_1, \mathcal{W}_2$ on $B \times E^3$ in two different situations: when $\partial \mathcal{C}$ is smooth and when $\partial \mathcal{C}$ is a polyhedron:

- In both cases, the evaluation of $\mathcal{M}_0$ gives

$$\mathcal{M}_0^{\mathcal{C}}(B) = < \mathbf{N}(\mathcal{C})_{|B \times \mathbb{E}^3}, \mathcal{W}_0 > = \mathcal{A}(B). \tag{24.9}$$

- The evaluation of $\mathcal{M}_1$ gives:

  - If the boundary of $\mathcal{C}$ is the smooth surface S,

$$\mathcal{M}_1^{\mathcal{C}}(B) = < \mathbf{N}(\mathcal{C})_{|B \times \mathbb{E}^3}, \mathcal{W}_1 > = \int_B H da. \tag{24.10}$$

– If the boundary of $\mathcal{C}$ is the triangulation T,

$$\mathcal{M}_1^{\mathcal{C}}(B) =< \mathbf{N}(\mathcal{C})_{|B \times \mathbb{E}^3}, \mathcal{W}_1 >= \sum_{e \text{ edge } \in B} l(e) \angle(e), \qquad (24.11)$$

where $\angle(e)$ denotes the angle between the normals of the two triangles incident to e, endowed with the suitable sign.

• The evaluation of $\mathcal{M}_2$ gives:

– If the boundary of $\mathcal{C}$ is the smooth surface S,

$$\mathcal{M}_2^{\mathcal{C}}(B) =< \mathbf{N}(\mathcal{C})_{|B \times \mathbb{E}^3}, \mathcal{W}_2 >= \int_B G da. \qquad (24.12)$$

– If the boundary of $\mathcal{C}$ is the triangulation T,

$$\mathcal{M}_2^{\mathcal{C}}(B) =< \mathbf{N}(\mathcal{C})_{|B \times \mathbb{E}^3}, \mathcal{W}_2 >= \sum_{v \text{ vertex } \in B} \sum_i \alpha_i(v), \qquad (24.13)$$

i.e., $\mathcal{M}_2^{\mathcal{C}}(B)$ is the sum over all vertices p lying in B of the angle defect at v.

Globally, by taking $B = S$ or T, one finds:

• If the boundary of $\mathcal{C}$ is the smooth surface S,

$$\mathcal{M}_0(\mathcal{C}) =< \mathbf{N}(\mathcal{C}), \mathcal{W}_0 >= \mathcal{A}(S),$$

$$\mathcal{M}_1(\mathcal{C}) =< \mathbf{N}(\mathcal{C}), \mathcal{W}_1 >= \int_S H da, \qquad (24.14)$$

$$\mathcal{M}_2(\mathcal{C}) =< \mathbf{N}(\mathcal{C}), \mathcal{W}_2 >= \int_S G da = \chi(S).$$

• If the boundary of $\mathcal{C}$ is the triangulation T,

$$\mathcal{M}_0(\mathcal{C}) =< \mathbf{N}(\mathcal{C}), \mathcal{W}_0 >= \mathcal{A}(T),$$

$$\mathcal{M}_1(\mathcal{C}) =< \mathbf{N}(\mathcal{C}), \mathcal{W}_1 >= \sum_{e \text{ edge } \subset T} l(e) \angle(e), \qquad (24.15)$$

where $\angle(e)$ denotes the angle between (the normals of) the two triangles incident to e, endowed with the suitable sign resulting from the additivity formula (20.1): the sign of $\angle(e)$ is positive if e is convex and negative if it is concave.

$$\mathcal{M}_2(\mathcal{C}) =< \mathbf{N}(\mathcal{C}), \mathcal{W}_2 >= \sum_{v \text{ vertex } \in T} \sum_i \alpha_i(v) = \chi(T), \qquad (24.16)$$

i.e., $\mathcal{M}_2(\mathcal{C})$ is the sum over all vertices v lying in T of the angle defect at v.

24.4 Second Fundamental Measure for Surfaces

The expression of the vector-valued form **h** (defined in Chap. 22) can be much simplified for surfaces in $\mathbb{E}^3$ by using the vector product. We still use the notation of Sect. 24.3.3. One has the following result (see [35] for detail).

Proposition 21. *Let X and Y be two (constant) vectors of* $\mathbb{E}^3$ *and let* (p,ξ) *be a point of* $\mathbb{E}^3 \times \mathbb{E}^3$. *Then*

$$\mathbf{h}^{X,Y}_{(p,\xi)} = (\xi \times X) \wedge Y. \tag{24.17}$$

In this proposition, $\times$ is the standard vector product in $\mathbb{E}^3$. Moreover, the vectors X and Y are identified with the 1-forms $< X,. >$ and $< Y,. >$, where $<,>$ denotes the scalar product of $\mathbb{E}^3$. As in the general case, the expression of the associated measure for a smooth surface S bounding a domain $\mathcal{C}$ is

$$\begin{aligned}
\mathbf{h}^{X,Y}_S(B) &= < \mathbf{N}(\mathcal{C}), \chi_{B \times \mathbb{E}^3} \mathbf{h}^{X,Y} > \\
&= \int_B h_S(\mathrm{pr}_{TS}X, \mathrm{pr}_{TS}Y)dv_S,
\end{aligned} \tag{24.18}$$

where B denotes a Borel subset of S and h_S denotes the second fundamental form of S.

The expression of this measure for a (triangulated) polyhedron T bounding a domain $\mathcal{K}$ of $\mathbb{E}^3$ is

$$\begin{aligned}
\mathbf{h}^{X,Y}_T(B) &= < \mathbf{N}(\mathcal{K}), \chi_{B \times \mathbb{E}^3} \mathbf{h}^{X,Y} > \\
&= \sum_{e \text{ edge of } T} \frac{l(e \cap B)}{2} [(\beta(e) - \sin \beta(e)) < e^+, X >< e^+, Y > \\
&\quad + (\beta(e) + \sin \beta(e)) < e^-, X >< e^-, Y >)],
\end{aligned} \tag{24.19}$$

where e^+ (resp., e^-) denotes the normalized sum (resp., difference) of the unit outward normal vector to the triangles incident to e and, as usual, $\beta(e)$ denotes the angle between these normal vectors. The sign of this angle is important for explicit computations: it is positive if the edge e is convex, and negative otherwise.

It must be remarked that $\mathbf{h}_T(B)(X,Y)$ is symmetric in (X,Y). This implies that $\mathbf{h}_T(B)$ can be diagonalized. One gets *generically* three different eigendirections: the one associated to the smallest eigenvalue can be called the *estimated normal vector field* of B and the two others can be called its *principal directions*, by analogy with the smooth case. By integrating these two directions, one gets curves on the polyhedron which can be called *lines of curvatures*. We insist on the fact that this last approach is *generic*: if the rank of $\mathbf{h}_T(B)$ is 1, one cannot detect a particular smallest eigendirection to define the normal vector field.

Remark When dealing with triangulations, it may be simpler to introduce a slightly different vector-valued form which gives simpler results: let us consider the form

$$\overline{\mathbf{h}}^{X,Y}_{(p,\xi)} = X \wedge (\xi \times Y). \tag{24.20}$$

Then, its evaluation on the normal cycle of a smooth surface S gives

$$\overline{\mathbf{h}}_S(B)(X,Y) = <\mathbf{N}(S), \chi_{(B \times \mathbb{E}^3} \overline{\mathbf{h}}^{X,Y}>$$
$$= \int_B h_S(j\mathrm{pr}_{TS}X, j\mathrm{pr}_{TM}Y)dv_S \tag{24.21}$$

for every Borel subset of $\mathbb{E}^3$, where h_S denotes the second fundamental form of S and j denotes the almost complex structure defined on each tangent plane of S. In other words, at any point m of S, if the matrix of h_S is

$$\begin{pmatrix} \lambda_1 & 0 \\ 0 & \lambda_2 \end{pmatrix}$$

in a suitable frame, then the matrix of $h_S(j., j.)$ written in the same frame is

$$\begin{pmatrix} \lambda_2 & 0 \\ 0 & \lambda_1 \end{pmatrix}.$$

This is equivalent replacing h_S by $(\mathrm{Trace}(h)\mathrm{Id} - h_M)$. Although this seems to be a little more complicated in the smooth case, it is simpler for polyhedra: if T is a (triangulated) polyhedron and B is a Borel subset of $\mathbb{E}^3$, then

$$h_T(B)(X,Y) = \sum_{e \text{ edge of } T} \beta(e)\mathrm{length}(e \cap B) < \overrightarrow{e}, X >< \overrightarrow{e}, Y > . \tag{24.22}$$

Chapter 25
Approximation of the Curvature of Curves

We have seen in Chap. 2 that the curvature of a smooth curve is the limit of sequences of the curvature of polygons inscribed in it, and tending to it with respect to the Hausdorff topology. We deal now with the problem of the *approximation* of the curvature of a smooth curve by the curvature of curves or polygons (not necessarily inscribed in it), close to it.

25.1 Curves in $\mathbb{E}^2$

Let us begin with the case of a curve in the (oriented) Euclidean plane $\mathbb{E}^2$. The problem is quite simple: let

$$\gamma : [0, l] \rightarrow \mathbb{E}^2$$
$$s \rightarrow \gamma(s)$$

be a regular curve (parametrized by the arc length). Let ε be a sufficiently small real number. We have seen in Chap. 2 that the map

$$\theta : [0, \varepsilon] \rightarrow \,]0, 2\pi[$$
$$s \rightarrow \theta(s) \tag{25.1}$$

(which associates to each $s \in [0, \varepsilon]$ the angle $\theta(s) \in [0, 2\pi[$ made by the (unit) tangent vector field $t(s)$ with the Ox-axis in $\mathbb{E}^2$) satisfies

$$\int_0^\varepsilon \kappa(s) ds = \theta_\varepsilon - \theta_0,$$

where κ is the *signed* curvature of the curve γ. An approximation of this integral is then equivalent to an approximation of these tangent vectors at the end points.

Globally, one can define the map

$$\tilde{\theta} : [0, l] \rightarrow \mathbb{R}$$
$$s \rightarrow \tilde{\theta}(s),$$

Fig. 25.1 The global *signed* curvature of a plane curve is the (algebraic) angle (on $\mathbb{S}^1$) between the tangent vectors at the end points. Consequently, a good approximation of t_0 and t_1 induces a good approximation of the global *signed* curvature

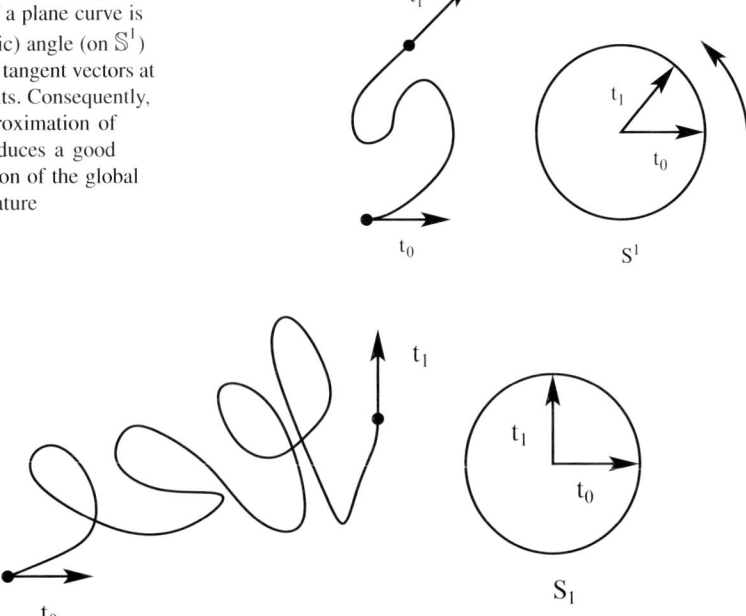

Fig. 25.2 The global *signed* curvature of this plane curve equals $6\pi + \frac{\pi}{2}$, since the rotation number equals 3. The approximation of the global *signed* curvature needs the knowledge of the rotation number of the curve

which satisfies

$$\int_0^l \kappa(s)ds = \tilde{\theta}_l - \tilde{\theta}_0,$$

where the angle $\tilde{\theta}$ can be written as

$$\tilde{\theta}(l) = 2\pi I + \delta,$$

I being the (algebraic) rotation number of the curve (see Sect. 2.4.3) plus a remainder δ. Consequently, one gets trivially an approximation of the global *signed* curvature of a plane curve γ if one knows the integer I and if one can approximate the tangent vectors at its end points (Figs. 25.1 and 25.2).

25.2 Curves in $\mathbb{E}^3$

Let us now consider a regular curve

$$\gamma : [0,l] \longrightarrow \mathbb{E}^3$$
$$s \longrightarrow \gamma(s),$$

parametrized by the arc length, and let

$$t : [0,l] \to \mathbb{S}^2$$

be its (unit) tangent vector field (here, we identify t with the Gauss map of γ). By definition,

$$\int_0^l k(s)ds = \int_0^l |t'(s)|ds.$$

In other words, the integral of the curvature k of γ is the length of the curve $t' \in \mathbb{S}^2 \subset \mathbb{E}^3$.

Suppose now that

$$c : [0,\bar{l}] \to \mathbb{E}^3$$

is a curve "close" to γ. Let us consider two situations:

1. If c is smooth and regular, it can be parametrized by the arc length. Let $\bar{t}$ be its unit tangent vector field identified with the Gauss map

$$\bar{t} : [0,\bar{l}] \to \mathbb{S}^2.$$

The integral of the curvature is the length of the curve $\bar{t}$. Let us consider the curves t and $\bar{t}$ as curves in $\mathbb{E}^3$. If $\bar{t}$ is closely near t, the general Theorem 43 can be written as

$$\frac{\cos \varpi_{max}}{1 + \omega_t(\bar{t})} \int_0^{\bar{l}} k(c) \leq \int_0^l k(\gamma) \leq \frac{\cos \varpi_{min}}{1 - \omega_t(\bar{t})} \int_0^{\bar{l}} k(c), \qquad (25.2)$$

where $\omega_t(\bar{t})$ is the relative curvature of the curve $\bar{t}$ with respect to the curve t and ϖ_{max} (resp., ϖ_{min}) denotes the maximal (resp., minimal) angular deviation of the curves $\bar{t}$ with respect to t (see Definition 34):

- Let us evaluate $\omega_t(\bar{t})$. By definition,

$$\omega_t(\bar{t}(u)) = |\overrightarrow{\mathrm{pr}(\bar{t}(u))\bar{t}(u)}| \mathbf{k}(\mathrm{pr}(\bar{t}(u)))$$

at each point $\bar{t}(u), u \in [0,\bar{l}]$ (where $\mathrm{pr}(\bar{t}(u))$ denotes the orthogonal projection onto the curve t of the point $\bar{t}(u)$ and $\mathbf{k}$ is the curvature function of the curve t). Moreover,

$$\omega_t(\bar{t}) = \sup_{u \in [0,\bar{l}]} \omega_t(\bar{t})(u).$$

Considering the curve t as a curve in $\mathbb{E}^3$, let us evaluate its curvature $\mathbf{k}$. We have

$$\mathbf{k} = \frac{|t' \wedge t''|}{|t'|^3}.$$

Since

$$t' = kn, \quad t'' = k'n - k\tau b + k^2 t$$

(where t, n, b are the Frnet frame of γ and τ is the torsion of γ), we deduce that

$$\mathbf{k} = (k^4 + \tau^2)^{\frac{1}{2}} k^{\frac{1}{3}}. \tag{25.3}$$

Then

$$\varpi_t(\bar{t}) = \sup_{u \in [0, \bar{l}]} (k^4 + \tau^2)^{\frac{1}{2}} k^{\frac{1}{3}}_{\mathrm{pr}(\bar{t}(u))} |\overrightarrow{\mathrm{pr}(\bar{t}(u))\bar{t}(u)}|. \tag{25.4}$$

- Let us evaluate the angular deviation function ϖ. At each point $\bar{t}(u)$, ϖ_u denotes the angle $\angle(t'_{\mathrm{pr}(\bar{t}(u))}, \bar{t}'_{t(u)})$, i.e., the angle between the two principal normals n_s and $n'(u)$, if s is the point of $[0, l]$ such that $\mathrm{pr}(\bar{t}(u)) = t(s)$.

2. Suppose now that γ is approximated by a polygonal line $P = v_1 v_2 ... v_n$. Let $e_1, ..., e_{n-1}$ be its (oriented) edges. As we have seen in Definition 5, the image of an edge e of P by its Gauss map $\mathbf{G}$ is the unit vector of $\mathbb{S}^2$ parallel to e (with the same orientation). If e_i and e_{i+1} are two consecutive edges on P, then the length of the arc of the great circle on $\mathbb{S}^2$ joining $\mathbf{G}(e_i)$ and $\mathbf{G}(e_{i+1})$ is the angle $\angle(e_i, e_{i+1})$. Consequently, the union of these arcs is a (continuous) curve C whose length is nothing but the sum of the angles between consecutive edges of P. By Definition 2, the curvature of P is the real number

$$\mathcal{K}(P) = \sum_{i=1}^{n-1} \angle(e_i, e_{i+1}). \tag{25.5}$$

Hence,

$$\mathcal{K}(P) = l(C). \tag{25.6}$$

Suppose that C is closely near t. Comparing $l(C)$ and $l(t)$ and still using (14.2), we get

$$\frac{\cos \varpi_{\max}}{1 + \varpi_t(C)} \mathcal{K}(P) \leq \int_0^l k(\gamma) \leq \frac{\cos \varpi_{\min}}{1 - \varpi_t(C)} \mathcal{K}(P), \tag{25.7}$$

where $\varpi_t(C)$ is the relative curvature of the curve $\bar{t}$ with respect to the curve t and $\varpi_{\max}$ (resp., $\varpi_{\min}$) denotes the maximum (resp., minimum) of the angular deviation function of the curves C with respect to t.

The effective evaluation of $\varpi_t(C)$, $\varpi_{\max}$, and $\varpi_{\min}$ is similar to the one studied in the first item, concerning the regular case.

Appendix: Comparing the Principal Normal Vectors

We summarize now approximation results which are often used in different problems concerning curves.

Fig. 25.3 β_{pq} is the angle between the tangent vectors at p and q of the curve γ. In this picture, the curve is planar

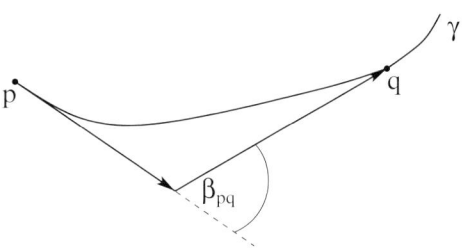

Comparing the Tangent Vectors of a Smooth Curve

Consider a smooth regular curve γ in $\mathbb{E}^N$. Proposition 22 bounds the angle between the tangent vectors of two points of γ. It is a direct application of the mean value theorem.

Proposition 22. *Let γ be a smooth regular oriented curve in $\mathbb{E}^N$, with end points p and q. Then, the angle $\beta_{pq} \in [0, \frac{\pi}{2}]$ between the tangent vectors at p and q of γ satisfies*

$$\sin \frac{\beta_{pq}}{2} \leq \frac{k_{\gamma_{\max}} l(\gamma)}{2}, \tag{25.8}$$

where $k_{\gamma_{\max}}$ is the maximal curvature of γ and $l(\gamma)$ is the length of γ (Fig. 25.3).

Denoting by t the unit tangent vector field of γ,

$$|t(q) - t(p)| \leq l(\gamma) \sup_{s \in [0,l]} |\frac{dt}{ds}|.$$

Since $|\frac{dt}{ds}| = k$, one deduces Proposition 22.

Comparing the Principal Normal Vectors of a Smooth Curve

Analogous bounds can be found if one compares the principal normal vectors at different points of a smooth curve. As in the previous section, it is a consequence of the mean value theorem. Consider a smooth regular curve γ in $\mathbb{E}^N$ with (unit) tangent vector field t. One can construct a *generalized Frénet frame* (see the Frénet equations (2.14) for curves in $\mathbb{E}^3$). The first vector fields of this frame satisfy

$$\frac{dt}{ds} = kn,$$
$$\frac{dn}{ds} = -kt + \hat{k}\hat{n}, \tag{25.9}$$

where $t, n, \hat{n}$ are orthonormal vector fields. The function $\hat{k}$ is called the *second curvature* and $\hat{n}$ is called the *second principal normal vector field* of γ.

Fig. 25.4 β_{pq} is the angle between the principal normal vectors at p and q of the curve γ. Here, the curve is planar

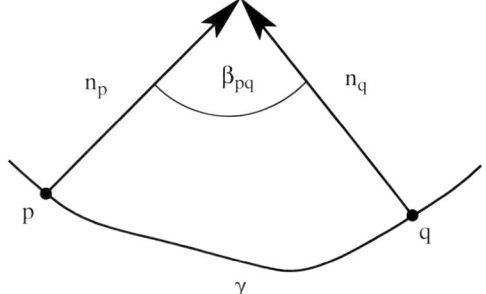

Theorem 75. *Let γ be a smooth regular oriented curve in $\mathbb{E}^N$, with end points p and q. The angle $\beta_{pq} \in [0, \frac{\pi}{2}]$ between the principal normals at p and q satisfies*

$$\sin(\beta_{pq}) \leq (\sqrt{k_{\max}^2 + \hat{k}_{\max}^2})l(\gamma), \tag{25.10}$$

where $k_{\max}$ is the maximal curvature of γ, $\hat{k}_{\max}$ is the maximal second curvature of γ, and $l(\gamma)$ is the length of γ (Fig. 25.4).

The proof is a direct consequence of the mean value theorem:

$$|n(b) - n(a)| \leq l(\gamma) \sup_{s \in [0,l]} |\frac{dn}{ds}|.$$

Using (25.9), one deduces (25.10).

If $N = 2$, $\hat{k}$ is null. If $N = 3$, $\hat{k}$ is the torsion τ of γ and $\hat{n}$ is the binormal vector field. Consequently, with the usual notation, we obtain the following result.

Corollary 17. *Let γ be a smooth regular oriented curve in $\mathbb{E}^N$, with end points p and q:*

1. If $N = 2$,

$$\sin(\beta_{pq}) \leq k_{\max}l(\gamma). \tag{25.11}$$

2. If $N = 3$,

$$\sin(\beta_{pq}) \leq (\sqrt{k_{\max}^2 + \tau_{\max}^2})l(\gamma). \tag{25.12}$$

A Bound Involving the Deviation Angle

We now deal with a curve approximated by its chord.

Theorem 76. *Let γ be a smooth regular oriented curve in $\mathbb{E}^N$, with end points p and q. The deviation angle $\alpha_{\max}$ of the chord pq with respect to γ satisfies*

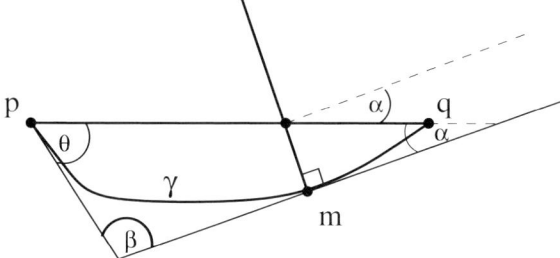

Fig. 25.5 The three angles α, β, and θ

$$\sin\frac{\alpha_{\max}}{2} \le \frac{3}{4}k_{\gamma\max}l(\gamma) \le \frac{3}{4}\frac{k_{\gamma\max}|\overrightarrow{pq}|}{1-\omega_\gamma(pq)}. \qquad (25.13)$$

Proof of Theorem 76 With the same notation, at each point m of γ, we have

$$
\begin{aligned}
2\sin\frac{\alpha_m}{2} &= |t_m - \frac{\overrightarrow{pq}}{|\overrightarrow{pq}|}| \\
&\le |t_m - t_p| + |t_p - \frac{\overrightarrow{pq}}{|\overrightarrow{pq}|}| \\
&= 2\sin\frac{\beta_{mp}}{2} + 2\sin\frac{\theta_p}{2} \\
&\le 2\sin\frac{\beta_{mp}}{2} + 2\sin\theta_p.
\end{aligned}
\qquad (25.14)
$$

Now, since the problem is symmetric in p and q, we can assume that the length of the curve between p and m is less than or equal to half the length of the curve γ. We deduce from (25.8) that

$$\sin\frac{\beta_{mp}}{2} \le \frac{k_{\gamma\max}l(\gamma)}{4}.$$

On the other hand, (14.4) implies that

$$\sin\theta_p \le \frac{k_{\gamma\max}l(\gamma)}{2}.$$

Since (25.14) is true for every point $m \in \gamma$, we deduce the left inequality of Theorem 76. The right one is deduced from (14.3) (Fig. 25.5). $\square$

Chapter 26
Approximation of the Curvatures of Surfaces

This chapter applies our general approximation Theorem 71 for surfaces in $\mathbb{E}^3$. In Sect. 26.1, we deal with a surface S approximated by another one, closely near it. In Sect. 26.2, we assume that the approximating surface is a triangulation. Finally, in Sect. 26.2.3, we assume that the triangulation is inscribed in S. We get sharper and sharper approximation and convergence results [34], [35].

26.1 The General Approximation Result

In Chaps. 21 and 22, we gave a general approximation result for the curvature measures and the second fundamental measure of a (smooth) hypersurface of $\mathbb{E}^N$ approximated by a geometric subset. This result depends inter alia on a good evaluation of the mass of the corresponding normal cycles. Using the notation of Chaps. 21 and 22 and simply putting $N = 3$ in the general Theorems 71 and 74, we get the corresponding result for surfaces in $\mathbb{E}^3$.

As usual, let us assume that $\mathcal{W}$ is a weakly regular geometric subset of $\mathbb{E}^3$, which bounds the geometric subset $\mathcal{K}$ of $\mathbb{E}^3$ in the sense of Definition 48. Moreover, assume that $\mathcal{W}$ is closely near a smooth (closed) surface S which bounds a compact domain K, as defined in Definition 33.

Since $N = 3$, the general theory allows us to define three curvature measures ($\mathcal{M}_0$, $\mathcal{M}_1$, and $\mathcal{M}_2$) and the second fundamental measure **h**:

- The curvature measure $\mathcal{M}_0$ is nothing but the area $\mathcal{A}$.
- The curvature measure $\mathcal{M}_1$ is the mean curvature measure.
- The curvature measure $\mathcal{M}_2$ is the Gauss curvature measure.

To be more intuitive, we put

$$\mathcal{M}_H = \mathcal{M}_1, \quad \mathcal{M}_G = \mathcal{M}_2.$$

The explicit expressions for $\mathcal{M}_H$ and $\mathcal{M}_G$ are given in Chap. 21. We already gave in Chap. 15 an approximation of the area of a surface. Let us focus our attention now to the approximation of the curvature measure $\mathcal{M}_H, \mathcal{M}_G$ and the second fundamental measure **h**.

Theorem 77.

$$|\mathcal{M}_H^{\mathcal{K}}(B) - \mathcal{M}_H^{K}(B)| \leq$$

$$(\delta_B + \alpha_B)[\frac{2(1 + |h_{\mathrm{pr}_W(B)}|)}{1 - \delta_B|h_{\mathrm{pr}_W(B)}|}]^2 (\mathbf{M}(\mathbf{N}(\mathcal{K})_{|T_B\mathbb{E}^3}) + \mathbf{M}(\partial\mathbf{N}(\mathcal{K}_{|T_B\mathbb{E}^3})));$$

$$|\mathcal{M}_G^{\mathcal{K}}(B) - \mathcal{M}_G^{K}(B)| \leq$$

$$(\delta_B + \alpha_B)[\frac{2(1 + |h_{\mathrm{pr}_W(B)}|)}{1 - \delta_B|h_{\mathrm{pr}_W(B)}|}]^2 (\mathbf{M}(\mathbf{N}(\mathcal{K})_{|T_B\mathbb{E}^3}) + \mathbf{M}(\partial\mathbf{N}(\mathcal{K}_{|T_B\mathbb{E}^3})));$$

$$|\mathbf{h}_{\mathcal{K}}(B) - \mathbf{h}_K(B)| \leq$$

$$(\delta_B + \alpha_B)[\frac{2(1 + |h_{\mathrm{pr}_W(B)}|)}{1 - \delta_B|h_{\mathrm{pr}_W(B)}|}]^2 (\mathbf{M}(\mathbf{N}(\mathcal{K})_{|T_B\mathbb{E}^3}) + \mathbf{M}(\partial\mathbf{N}(\mathcal{K}_{|T_B\mathbb{E}^3}))).$$

26.2 Approximation by a Triangulation

Theorem 77 is interesting if one can bound the involved normal cycles. If the approximating surface is a (closed oriented) triangulation T bounding a domain C in $\mathbb{E}^3$, this is quite easy. It is the goal of Sect. 26.2.1.

26.2.1 A Bound on the Mass of the Normal Cycle

The following (local) theorem shows that one can bound this mass by using the natural invariants involved in the geometry of the triangulation, together with the angular deviation. We need some notations. As before, B denotes the interior of a union of triangles of T. Moreover:

- α_B denotes as usual the angular deviation between B and $\mathrm{pr}(B)$.
- l_B denotes the sum of the lengths of the edges of T lying in B.
- $\mathbf{n}^e$ denotes the maximum number of edges having a common vertex in B.
- $\mathbf{n}^v$ denotes the number of vertices of T lying in B.
- $l_{\partial B}$ denotes the length of ∂B.
- $\mathbf{n}^{v_\partial}$ denotes the number of vertices of ∂B.

In addition, we use the notation $\mathbf{N}^e, \mathbf{N}^v$ defined in Sect. 24.3.3.

The following theorem gives a bound on the mass of the normal cycle of a triangulation.

Theorem 78. *The mass of the normal cycle of $\mathcal{C}$ is bounded as follows:*

1. $\mathbf{M}(\mathbf{N}^e(\mathcal{C})_{|B\times\mathbb{E}^3}) \leq 2\alpha_B l_B$.
2. $\mathbf{M}(\mathbf{N}^v(\mathcal{C})_{|B\times\mathbb{E}^3}) \leq 4\pi\sin^2\frac{\alpha_B}{2} l_B\mathbf{n}^v\mathbf{n}^e$.
3. $\mathbf{M}(\mathbf{N}(\mathcal{C})_{|B\times\mathbb{E}^3}) \leq \mathcal{A}(B) + 2\alpha_B l_B + 4\pi\sin^2\frac{\alpha_B}{2} l_B\mathbf{n}^v\mathbf{n}^e$.
4. $\mathbf{M}(\partial(\mathbf{N}(\mathcal{C})_{|B\times\mathbb{E}^3})) \leq l_{\partial B} + 2\alpha_B\mathbf{n}^{v_\partial}$.

Proof of Theorem 78

1. Over an edge e, the support of the normal cycle is reduced to a portion of cylinder, which can be identified with the product of e by an arc of circle c of the 2-sphere $\mathbb{S}^2$. The angle spanned by any point of c with the normal ξ of the surface at any vertex of e is smaller than α_B. The result follows by summing over all the edges of B.

2. Over a vertex v, the support of the normal cycle lies in $v\times\mathbb{S}^2$. We bound its mass as follows: consider two adjacent faces belonging to the 1-ring of v. The normals of these two faces span with ξ_v a geodesic triangle in $\mathbb{S}^2$ whose area is smaller than $4\pi\sin^2\frac{\alpha_B}{2}$. Moreover, the mass of the normal cycle over v is smaller than the sum of the areas of these geodesic triangles. Since the 1-ring of v contains a number of edges less than or equal to $\mathbf{n}^e$, the mass of the normal cycle over v is smaller than $\mathbf{n}^e\alpha_B^2$. Now, since B contains $\mathbf{n}^v$ vertices,

$$\mathbf{M}(\mathbf{N}^v(\mathcal{C})_{|B\times\mathbb{E}^3}) \leq 4\pi\sin^2\frac{\alpha_B}{2} l_B\mathbf{n}^v\mathbf{n}^e.$$

3. Since the normal cycle over a face can be identified with the face itself, one gets part (3) by summing the terms over the faces, the edges, and the vertices.

4. The boundary of $\mathbf{N}(\mathcal{C})_{|B\times\mathbb{E}^3}$ is composed of edges corresponding to the edges of ∂B, and arcs of circles above the vertices belonging to ∂B. We deduce part (4).
$\qquad\qquad\qquad\qquad\qquad\qquad\qquad\qquad\qquad\qquad\qquad\qquad\qquad\qquad\qquad\square$

26.2.2 Approximation of the Curvatures

We can now approximate the curvature measures of S with the curvature measures of T, as an immediate consequence of Theorem 67.

Theorem 79. *Let K be a compact subset of $\mathbb{E}^3$ whose boundary is a smooth (closed oriented embedded) hypersurface S. Let $\mathcal{C}$ be a compact subset of $\mathbb{E}^3$ whose boundary is a triangulation T closely near S. Let B be the interior of a union of triangles of T. Then,*

$$|\mathcal{M}_H^{\mathcal{C}}(B) - \mathcal{M}_H^K(\mathrm{pr}(B))| \leq$$

$$2\max(\delta_B,\alpha_B)(\frac{\sup_B(1,|h_B|)}{1-\delta_B|h_B|})^2(\mathcal{A}(B) + 2\alpha_B(l_B + \mathbf{n}^{v\delta}) + 4\pi\sin^2\frac{\alpha_B}{2}\mathbf{n}^e + l_{\partial B});$$

$$|\mathcal{M}_G^C(B) - \mathcal{M}_G^K(\mathrm{pr}(B))| \le$$

$$\max(\delta_B, \alpha_B)\left(\frac{\sup_B(1, |h_B|)}{1 - \delta_B|h_B|}\right)^2 (\mathcal{A}(B) + 2\alpha_B(l_B + \mathbf{n}^{\nu\delta}) + 4\pi \sin^2 \frac{\alpha_B}{2} \mathbf{n}^e + l_{\partial B});$$

$$|\mathbf{h}^C(B) - \mathbf{h}^K(\mathrm{pr}(B))| \le$$

$$\max(\delta_B, \alpha_B)\left(\frac{\sup_B(1, |h_B|)}{1 - \delta_B|h_B|}\right)^2 (\mathcal{A}(B) + 2\alpha_B(l_B + \mathbf{n}^{\nu\delta}) + 4\pi \sin^2 \frac{\alpha_B}{2} \mathbf{n}^e + l_{\partial B}).$$

26.2.3 Triangulations Closely Inscribed in a Surface

Let us now make a stronger assumption: the triangulation is *closely inscribed* in the surface S. We have seen in Chap. 15 that, under this assumption, one can control the angular deviation in terms of the shape of the triangles of T. Let us give a simpler version of Theorem 79, introducing the circumradius $r(t)$ of each triangle t of the triangulation.

Theorem 80. *Under the assumption of Theorem* 79:

- $|\mathcal{M}_H^C(B) - \mathcal{M}_H^K(\mathrm{pr}(B))| \le C_S \mathbf{K}\varepsilon,$
- $|\mathcal{M}_G^C(B) - \mathcal{M}_G^K(\mathrm{pr}(B))| \le C_S \mathbf{K}\varepsilon,$
- $|\mathbf{h}^C(B) - {}^K(\mathrm{pr}(B))| \le C_S \mathbf{K}\varepsilon,$

where:

- C_S *is a real number depending only on the maximum curvature of S.*
- $\mathbf{K} = \sum_{t \in \mathbf{T}, t \subset \overline{B}} r(t)^2 + \sum_{t \in \mathbf{T}, t \subset \overline{B}, t \cap \partial B \neq \emptyset} r(t).$
- $\varepsilon = \max\{r(t), t \in \mathbf{T}, t \subset \overline{B}\}.$

We will see in Chap. 27 that, in several cases, the number $\mathbf{K}$ can be bounded from above, implying that the curvature measures of a sequence of increasingly fine triangulations of a smooth surface converge to those of the smooth surface.

Remark These theorems can be interpreted as follows: suppose that one deals with a triangulated mesh T. This mesh can be considered as the approximation of an infinity of smooth surfaces S. Although it is in general impossible to evaluate the geometry of S without other assumptions, Theorems 79 and 80 claim that every smooth surface S, in which T is closely inscribed and

- whose normal vector field is close to the normal of the faces,
- whose second fundamental form is "not too big,"

has a local geometry close to that of T. Moreover, the error between the mutual curvatures is bounded by an explicit constant depending on the *intrinsic* geometry of T and on the two previous assumptions.

Chapter 27
On Restricted Delaunay Triangulations

We deal here with particular approximations of smooth curves or surfaces in $\mathbb{E}^N$ ($N = 2$ or $N = 3$): those which arise from Voronoi diagrams and Delaunay triangulations of $\mathbb{E}^N$. Although everything can be done in any dimension, we restrict ourself to $\mathbb{E}^2$ and $\mathbb{E}^3$ for simplicity. We refer to [16, 18, 19, 38] for a large and deep study of Voronoi diagrams, Delaunay triangulations, and more generally for surface reconstruction.

27.1 Delaunay Triangulation

27.1.1 Main Definitions

A *sample* $\mathcal{S}$ denotes simply a subset of points of $\mathbb{E}^N$ (which can be finite or infinite). In this section, we assume that the points of $\mathcal{S}$ are finite and *in general position*, i.e.:

- If $N = 2$, no subset of four points of $\mathcal{S}$ lies on a same circle of $\mathbb{E}^2$.
- If $N = 3$, no subset of five points of $\mathcal{S}$ lies on a same sphere of $\mathbb{E}^3$.

We defined the notion of triangulation in Chap. 6. We improve it a little by defining a triangulation *associated to a finite set of points*.

Definition 55. Let $\mathcal{S}$ be a finite set of points of $\mathbb{E}^N$. A *triangulation* of $\mathcal{S}$ is a simplicial cell complex embedded in $\mathbb{E}^N$ whose set of vertices is $\mathcal{S}$, and such that the union of its cells is the convex hull $\mathrm{conv}(\mathcal{S})$ of $\mathcal{S}$.

Using Sect. 4.1, we can define the Voronoi diagram $\mathrm{Vor}(\mathcal{S})$ associated to any sample $\mathcal{S}$ of $\mathbb{E}^N$, as a cell decomposition of $\mathbb{E}^N$ into convex tetrahedra (or triangles if $N = 2$). The Delaunay triangulation associated to the Voronoi diagram of $\mathcal{S}$ is a particular triangulation of $\mathcal{S}$, *dual* to $\mathrm{Vor}(\mathcal{S})$.

Recall the classical definition of duality between cell complexes: two cell complexes V and D are *dual* if there exists an *involutive correspondence* between the faces of V and the faces of D that reverses the inclusions, i.e., for any two faces f and g of V, their dual faces f^* and g^* of D satisfy

$$f \subset g \Longrightarrow g^* \subset f^*.$$

Suppose for instance that $N = 3$. If f is a face (of any dimension $k(0 \leq k \leq 3)$) of the Voronoi diagram of a sample $\mathcal{S}$, all points of the interior of f have the same closest points in $\mathcal{S}$. Let $\mathcal{S}_f \subset \mathcal{S}$ denote the subset of those closest points. The face dual to f is the convex hull of $\mathcal{S}_f$. Its dimension is $3 - k$.

Definition 56. The *Delaunay triangulation Del($\mathcal{S}$)* of $\mathcal{S}$ is the simplicial complex consisting of all the faces dual to Vor($\mathcal{S}$).

If we restrict our attention to $\mathbb{E}^3$, we can simply claim that the Delaunay triangulation associated to $\mathcal{S}$ is the simplicial complex whose vertices are the points of $\mathcal{S}$ and which decomposes the convex hull of $\mathcal{S}$ as follows: the convex hull of four points of $\mathcal{S}$ defines a three-dimensional cell if the intersections of the corresponding Voronoi cells are nonempty.

As before, one defines the Delaunay tetrahedra, Delaunay faces (or triangles), and Delaunay edges. The Delaunay vertices are nothing but the points of $\mathcal{S}$.

If $\mathcal{S}$ is in *general position*, then Del($\mathcal{S}$) is a triangulation of $\mathcal{S}$ in the sense of Definition 55.

The reader may adapt these considerations if the sample $\mathcal{S}$ lies in $\mathbb{E}^2$.

27.1.2 The Empty Ball Property

We still assume that $\mathcal{S}$ lies in $\mathbb{E}^3$. As an obvious consequence of the definition, the (relative) interior of a Voronoi k-face f is the set of points having exactly $3 - k + 1$ nearest points of the sample $\mathcal{S}$. Consequently, there exists a ball empty of points of $\mathcal{S}$, whose boundary is a sphere containing the vertices of the simplex f^* dual to f. One says that the simplex f^* has the *empty ball property*. Any Delaunay tetrahedron corresponds a *unique* empty ball. However, there is a continuous family of empty balls corresponding to a Delaunay face (triangle) or Delaunay edge. The following proposition characterizes the Delaunay triangulations in terms of the *empty ball property*.

Proposition 23. *A triangulation T of a finite set of points $\mathcal{S}$ is a Delaunay triangulation of $\mathcal{S}$ if any 3-simplex of T has a circumscribing 2-sphere that does not enclose any point of $\mathcal{S}$. Any 3-simplex with vertices in $\mathcal{S}$ which can be circumscribed by a 2-sphere that does not enclose any point of $\mathcal{S}$ is a face of a Delaunay triangulation of $\mathcal{S}$.*

As before, the reader may adapt these considerations if the sample $\mathcal{S}$ lies in $\mathbb{E}^2$.

27.1.3 Delaunay Triangulation Restricted to a Subset

Let $\mathcal{S}$ be a (finite) sample of $\mathbb{E}^3$ and let X be any subset of $\mathbb{E}^3$.

Definition 57.

1. If f is any k-face of $\mathrm{Vor}(\mathcal{S})$, $f \cap X$ is called the restriction of f to X.
2. The subcomplex $\mathrm{Vor}_{|X}(\mathcal{S})$ of all nonempty restrictions of faces of $\mathrm{Vor}(\mathcal{S})$ to X is called the restriction of $\mathrm{Vor}(\mathcal{S})$ to X.
3. The restriction to X of the Delaunay triangulation $\mathrm{Del}(\mathcal{S})$ is the subcomplex $\mathrm{Del}_X(\mathcal{S})$ of $\mathrm{Del}(\mathcal{S})$, which is the union of the k-faces of $\mathrm{Del}(\mathcal{S})$ $(0 \leq k \leq 2)$ whose dual Voronoi $(3-k)$-faces intersect X.

The main problem is to know if this new triangulation $\mathrm{Del}_X(\mathcal{S})$ is a "good approximation" of X (in a sense to be specified). This is the objective of Sect. 27.2. It will essentially depend on the position of $\mathcal{S}$ with respect to X. Let us restrict our attention to the case of surfaces (Figs. 27.1 and 27.2).

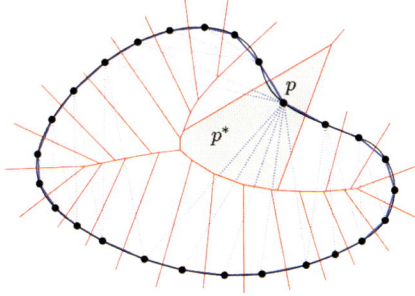

Fig. 27.1 Delaunay triangulation of a point set sampling a smooth curve. Edges of the restricted triangulation are shown as *solid blue lines*, other Delaunay edges of the Voronoi complex are shown as *red lines*. This image is courtesy of Steve Oudot, I.N.R.I.A. Geometrica

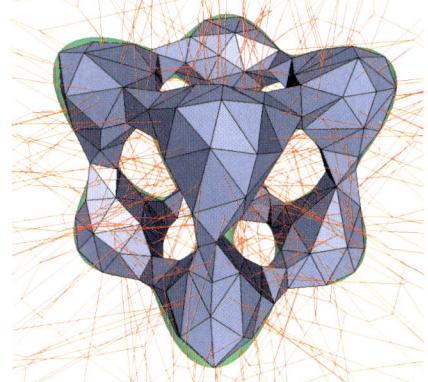

Fig. 27.2 Restriction (*in blue*) of a Delaunay triangulation to a smooth algebraic surface (*in green*). Voronoi edges are shown in *red*. This image is courtesy of Steve Oudot, I.N.R.I.A. Geometrica

27.2 Approximation Using a Delaunay Triangulation

As we have seen in Chap. 26, the approximation of the geometric invariants of a surface by another close surface is "good" if the corresponding normals are "close." If a smooth surface S is approximated by a Delaunay triangulation restricted to it, as constructed in the previous sections, we must ensure that the normals of the faces of the triangles are close to the normals of the surface S (at the corresponding points).

27.2.1 The Notion of ε-Sample

In a series of papers, Amenta et al. [5, 7] gave sufficient conditions for a sample of points on a surface to be *interesting*, in terms of the *local feature size* lfs defined in Chap. 4. Let us summarize them.

Definition 58. Let ε be a real number such that $0 < \varepsilon < 1$. A set of points $\mathcal{S}$ on X is an ε-sample of X if and only if, for every point m of X, the ball $B(m, \varepsilon \text{lfs}(m))$ encloses at least one point of $\mathcal{S}$.[1]

An ε-sample of X will generally be denoted by $\mathcal{S}_\varepsilon$.

27.2.2 A Bound on the Hausdorff Distance

27.2.2.1 The Case of a Smooth Curve of $\mathbb{E}^N$

Let c be a smooth (connected) curve of $\mathbb{E}^N$ and let $\mathcal{S}$ be a sample of c. We say that $\mathcal{S}$ is in *natural position* with respect to c if it satisfies the following (natural) conditions:

- $\mathcal{S}$ contains at least three points and two of them are the end points of c.
- No vertex for Vor $(\mathcal{S})$ lies on $\mathcal{S}$.

If an ε-sample $\mathcal{S}_\varepsilon$ of a curve c is in natural position, then $\text{Del}_c(\mathcal{S}_\varepsilon)$ tends to 0 with ε. More precisely, we have the following result.

Theorem 81. *Let c be a smooth (connected) curve of $\mathbb{E}^N$ and let $\mathcal{S}_\varepsilon$ be an ε-sample of c in natural position. Then, $\text{Del}_c(\mathcal{S}_\varepsilon)$ is a polygonal curve homeomorphic to c, and the Hausdorff distance between c and $\text{Del}_c(\mathcal{S}_\varepsilon)$ satisfies*

$$\text{d}(\text{Del}_c(\mathcal{S}_\varepsilon), c) \le 2\varepsilon \sup_{m \in c} \text{lfs}(m).$$

The proof of this theorem, its improvements, and related results can be found in [6, 16, 39].

---

[1] The reader must be careful: some authors replace $B(m, \varepsilon \text{lfs}(m))$ by $B(m, \varepsilon)$ in Definition 58, which implies confusion in many theorems. We will only use Definition 58 in this chapter.

27.2.2.2 The Case of a Smooth Surface

The previous result can be extended to surfaces in $\mathbb{E}^3$. Boissonnat and Oudot [17] proved the following theorem.

Theorem 82. *Let S be a smooth surface in $\mathbb{E}^3$ and $\mathcal{S}_\varepsilon$ be an ε-sample of S. Then, if $\varepsilon < 0.18$:*

1. *$Del_S(\mathcal{S}_\varepsilon)$ is closely inscribed in S (and consequently, S and $\mathcal{S}_\varepsilon$ are homeomorphic).*
2. *Moreover, the Hausdorff distance between $Del_S(\mathcal{S}_\varepsilon)$ and S satisfies^2*

$$\mathrm{d}(Del_S(\mathcal{S}_\varepsilon), S) \le 4.5\varepsilon^2 \sup_{m \in S} \mathrm{lfs}(m).$$

27.2.3 Convergence of the Normals

The following result shows that if ε is smooth enough, then the restricted Delaunay triangulation associated to the sample S is "in good position."

27.2.3.1 The Case of a Smooth Curve

For curves, the following classical result (see for instance [38, Lemma 2.4, p. 29]) shows that the angular deviation between a smooth curve c and $Del_c(\mathcal{S}_\varepsilon)$ tends to 0 with ε.

Theorem 83. *Let c be a smooth (connected) curve of $\mathbb{E}^2$ and let $\mathcal{S}_\varepsilon$ be an ε-sample of c in natural position. Then:*

1. *The angular deviation $\alpha_{\max}$ of $Del_c(\mathcal{S}_\varepsilon)$ with respect to c satisfies*

$$\sin \alpha_{\max} \le \frac{\varepsilon}{1 - \varepsilon}.$$

2. *In particular,*
$$\alpha_{\max} = O(\varepsilon).$$

27.2.3.2 The Case of a Smooth Surface

Theorem 83 can be extended to surfaces [5, 38].

---

^{2} There is a slight difference between Theorem 82 and Theorem 4.5 of [17], in which ε must be smaller than 0.091. This is due to the fact that the authors introduce and deal with the concept of *loose ε-sample*, weaker than the usual one.

Theorem 84. *Let S_ε be an ε-sample on S. Then, the angular deviation $\alpha_{\max}$ of $Del_S(S_\varepsilon)$ with respect to S satisfies*

$$\alpha_{\max} = O(\varepsilon).$$

27.2.4 Convergence of Length and Area

By definition, the length of a smooth curve is the supremum of the lengths of the polygonal lines closely inscribed in it. We have seen many times that this result has no immediate generalization to surfaces (see the *Lantern of Schwarz* in Chap. 3). However, if we restrict our attention to ε-samples, we have easy convergence theorems: as an immediate consequence of Theorem 84, we get the following result.

Theorem 85. *Let S be a closed surface of $\mathbb{E}^3$. Let $Del_S(S_\varepsilon)$ be the restricted Delaunay triangulation of an ε-sample S_ε with respect to S. Then*

$$\lim_{\varepsilon \to 0} \mathcal{A}(Del_S(S_\varepsilon)) = \mathcal{A}(S).$$

27.2.5 Convergence of Curvatures

Unfortunately, we cannot get further results on the convergence of the curvature measures of $Del_S(S_\varepsilon)$ to those of S without additional assumptions. In fact, to apply Theorem 66 for instance, we need to bound the mass of the normal cycle $Del_S(S_\varepsilon)$. In general, this mass is not bounded when ε tends to 0 (see [35] for a counterexample). That is why we need a stronger assumption on the sample. As an example, one can use sequences of κ-light samples.

Definition 59. *An ε-sample S_ε on a surface S is κ-light if, for every point $m \in S$, the ball $B(m, \varepsilon lfs(m))$ encloses at most κ-points of S (Fig. 27.3).*

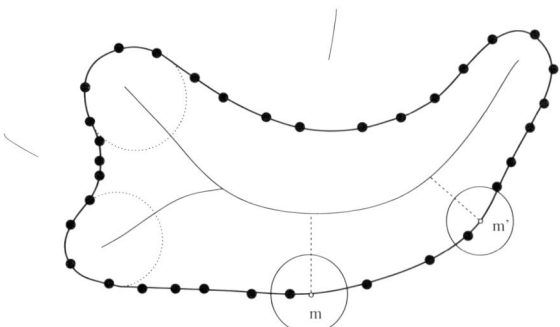

Fig. 27.3 An ε-sample S of a closed curve in $\mathbb{E}^2$, where $\varepsilon = \frac{1}{2}$: for each point m of the curve, the circle of center m and radius $\frac{1}{2} lfs(m)$ contains at least one point of S

Corollary 18. *Let $Del_S(\mathcal{S}_\varepsilon)$ be the restricted Delaunay triangulation of a κ-light ε-sample of a smooth surface S, where κ is a positive real number. Let $\widetilde{B}$ be a Borel subset of $\mathbb{E}^N$ and let $B_\varepsilon = \widetilde{B} \cap Del_S(\mathcal{S}_\varepsilon)$. Then*

$$
\begin{aligned}
|\mathcal{M}_G^{Del_S(\mathcal{S}_\varepsilon)}(B_\varepsilon) - \mathcal{M}_G^S(\mathrm{pr}_S(B_\varepsilon))| &= O(\varepsilon), \\
|\mathcal{M}_H^{Del_S(\mathcal{S}_\varepsilon)}(B_\varepsilon) - \mathcal{M}_H^S(\mathrm{pr}_S(B_\varepsilon))| &= O(\varepsilon), \\
|\tilde{h}^{Del_S(\mathcal{S}_\varepsilon)}(B_\varepsilon) - \tilde{h}^S(\mathrm{pr}_S(B_\varepsilon))| &= O(\varepsilon).
\end{aligned}
\tag{27.1}
$$

We leave the precise proof to the reader.

Bibliography

1. A.D. Aleksandrov, *The existence almost everywhere of the second differential of a convex function and some associated properties of surfaces*, Ucenye Zapiski Leningr. Gos. Univ. Ser. Mat., 37 (1939) 3–35
2. A.D. Aleksandrov, V.A. Zalgaller, *Intrinsic Geometry of Surfaces*, Translations of Mathematical Monographs, vol. 15, 1967
3. C.B. Allendoerfer, A. Weil, *The Gauss–Bonnet theorem for Riemannian polyhedra*, Trans. Am. Math. Soc. 53 (1943) 101–129
4. L. Ambrosio, N. Fusco, D. Pallara, *Functions of Bounded Variation and Free Discontinuity Problems*, Clarendon, Oxford, 2006
5. N. Amenta, M. Bern, *Surface reconstruction by Voronoi filtering*, Discrete Comput. Geom. 22 (1999) 481–504
6. N. Amenta, M. Bern, D. Eppstein, *The crust and the β-skeleton: Combinatorial curve reconstruction*, Graph. Models Image Process. 60 (1998) 125–135
7. N. Amenta, S. Choi, T.K. Dey, N. Leekha, *A simple algorithm for homeomorphic surface reconstruction*, 16th Annual Symposium on Computational Geometry, 2000, pp. 213–222
8. Th. Banchoff, *Critical points and curvature for embedded polyhedra*, J. Diff. Geom. 1 (1967) 245–256
9. M. Berger, *Géométrie*, vol. 2 and 3, Cedic/Fernand Nathan, Paris, 1978
10. M. Berger, *A Panoramic View of Riemannian Geometry*, Springer, Berlin Heidelberg New York, 2003
11. M. Berger, B. Gostiaux, *Géométrie différentielle: variétés, courbes et surfaces*, Presses Universitaires de France, Paris, 1993
12. A. Bernig, *The normal cycle of compact definable sets*, Israel J. Math. 159 (2007) 373–411
13. A. Bernig, L. Bröcker, *Lipschitz–Killing invariants*, Math. Nachr. 245 (2002) 5–25
14. A. Bernig, L. Bröcker, *Courbures intrinsèques dans les catégories analytico-géométriques*, Ann. Inst. Fourier 53 (2003) 1897–1924
15. A. Bernig, L. Bröcker, *Curvature tensors of singular spaces*, Diff. Geom. Appl. 24 (2006) 191–208
16. J.D. Boissonnat, *Voronoi Diagrams, Triangulations and Surfaces*, Cours de $M2$, 2006
17. J.D. Boissonnat, S. Oudot, *Provably good sampling and meshing of surfaces*, Graph. Models 67(5) (2005) 405–451
18. J.D. Boissonnat, M. Teillaud (eds.), *Effective Computational Geometry for Curves and Surfaces*, Mathematics and Visualization, Springer, Berlin Heidelberg New York, 2007
19. J.D. Boissonnat, M. Yvinnec, *Algorithmic Geometry*, Cambridge University Press, Cambridge, 1998
20. V. Borrelli, *Courbures discrètes*, mémoire de DEA de l'Université Claude Bernard, France, 1993

21. V. Borrelli, F. Cazals, J.M. Morvan, *On the angular defect of triangulations and the pointwise approximation of curvatures*, Comput. Aided Geom. Des. 20 (2003) 319–341
22. U. Brehm, W. Kühnel, *Smooth approximation of polyhedral surfaces regarding curvatures*, Geom. Dedicata 12 (1982) 435–461
23. E. Calabi, P. Olver, A. Tannenbaum, *Affine geometry, curve flows and invariant numerical approximations*, Adv. Math. 124 (1996) 154–196
24. J. Cheeger, W. Müller, R. Schrader, *On the curvature of piecewise flat spaces*, Commun. Math. Phys. 92 (1984) 405–454
25. J. Cheeger, W. Müller, R. Schrader, *Kinematic and tube formulas for piecewise linear spaces*, Indiana Univ. Math. J. 35 (1986) 737–754
26. B.Y. Chen, *Geometry of Submanifolds*, Dekker, New York, 1973
27. B.Y. Chen, *Total Mean Curvature and Submanifolds of Finite Type*, Series in Pure Mathematics, vol. 1, World Scientific, Singapore, 1984
28. S.S. Chern, *A simple intrinsic proof of the Gauss–Bonnet formula for closed Riemannian manifolds*, Ann. Math. 45 (1944) 747–752
29. S.S. Chern, *On the curvature integral in a Riemannian manifold*, Ann. Math. 46 (1945) 674–684
30. S.S. Chern, *On the kinematic formula in the Euclidean space of N dimensions*, Am. J. Math. 74 (1952) 227–236
31. S.S. Chern, R.K. Lashof, *On the total curvature of immersed manifolds*, Am. J. Math. 79 (1957) 306–318
32. S.S. Chern, R.K. Lashof, *On the total curvature of immersed manifolds II*, Mich. Math. J. 5 (1958) 5–12
33. D. Cohen-Steiner, H. Edelsbrunner, *Inequalities for the curvature of curves and surfaces*, Proc. 21th ACM Symposium on Computational Geometry, 2005, pp. 272–277
34. D. Cohen-Steiner, J.-M. Morvan, *Restricted Delaunay triangulations and normal cycle*, Proceedings of the 19th Annual Symposium on Computational Geometry, 2003, pp. 312–321
35. D. Cohen-Steiner, J.-M. Morvan, *Second fundamental measure of geometric sets and local approximation of curvatures*, J. Diff. Geom. 74 (2006) 363–394
36. M. Craizer, Th. Lewiner, J.M. Morvan, *Parabolic polygons and discrete affine geometry*, Sibgrapi (19th Brazilian Symposium on Computer Graphics and Image Processing), 2006, pp. 19–26
37. M.C. Delfour, J.-P. Zolésio, *Shape Analysis via Oriented Distance Functions*, J. Funct. Anal. 123(1) (1994) 129–201
38. T.K. Dey, *Curve and Surface Reconstruction: Algorithms with Mathematical Analysis*, Cambridge University Press, Cambridge, 2006
39. T.K. Dey, P. Kumar, *A simple provable curve reconstruction algorithm*, Proceedings of the 10th Annual ACM–SIAM Symposium on Discrete Algorithms, 1999, pp. 893–894
40. M.P. Do Carmo, *Differential geometry of curves and surfaces*, Prentice-Hall, Englewood Cliffs, 1976 (translated from the Portuguese)
41. I. Fary, *Sur la courbure totale d'une courbe gauche faisant un noeud*, Bull. Soc. Math. France 77 (1949) 128–138
42. H. Federer, *Curvature measures*, Trans. Am. Math. Soc. 93 (1959) 418–491
43. H. Federer, *Geometric Measure Theory*, Springer, Berlin Heidelberg New York, 1983
44. W. Fenchel, *Über Krümmung und Windung geschlossener Raumkurven*, Math. Ann. 101 (1929) 238–252
45. M. Fréchet, *Sur quelques points du calcul fonctionnel*, Rendiconti del Circolo Mathematico di Palermo 22 (1906) 1–74
46. J. Fu, *Curvature measures and generalized Morse theory*, J. Diff. Geom. 30 (1989) 619–642
47. J. Fu, *Monge–Ampère functions I*, Indiana Univ. Math. J. 38 (1989) 745–771
48. J. Fu, *Convergence of curvatures in secant approximations*, J. Diff. Geom. 37 (1993) 177–190
49. J. Fu, *Curvature measures of subanalytic sets*, Am. J. Math 116 (1994) 819–880
50. J. Fu, *Curvature of singular spaces via the normal cycle*, Am. Math. Soc. 116 (1994) 819–880
51. W. Greub, S. Halperin, R. Vanstone, *Connections, Curvature, and Cohomology*, vol. 1–3, Academic, New York, 1972

52. H. Hadwiger, *Vorlesungen über Inhalt, Oberfläche, Isoperimetrie*, Springer, Berlin Heidelberg New York, 1957
53. P.R. Halmos, *Measure Theory*, Graduate Text in Mathematics, Springer, Berlin Heidelberg New York, 1950
54. M. Kashiwara, P. Schapira, *Sheaves on Manifolds*, Springer, Berlin Heidelberg New York, 1990
55. D.A. Klain, *A short proof of Hadwiger characterisation theorem*, Mathematika 42 (1995) 329–339
56. D.A. Klain, G.-C. Rota, *Introduction to Geometric Probability*, Accademia Nazionale dei Lincei, Cambridge University Press, Cambridge, 1999
57. S. Kobayashi, K. Nomizu, *Foundations of Differential Geometry*, vol. 1 and 2, Interscience, New York, 1963
58. S. Lang, *Algebra*, Addison-Wesley, Reading, MA, 1965
59. A. Lichnérowicz, *Calcul tensoriel*, Collection Armand Colin, Paris
60. J.W. Milnor, *On the total curvature of knots*, Ann. Math. 52 (1950) 248–257
61. J.W. Milnor, *Morse Theory*, Princeton University Press, Princeton, NJ, 1963
62. J.W. Milnor, J.D. Stasheff, *Characteristic Classes*, Princeton University Press, Princeton, NJ, 1974
63. F. Morgan, *Geometric Measure Theory*, Academic, New York, 1987
64. M. Morse, *The Calculus of Variations in the Large*, vol. 18, American Mathematical Society Colloquium, New York, 1934
65. J.M. Morvan, B. Thibert, *On the approximation of a smooth surface with a triangulated mesh*, Comput. Geom. Theory Appl. 23(3) (2002) 337–352
66. J.M. Morvan, B. Thibert, *Approximation of the normal vector field and the area of a smooth surface*, Discrete Comput. Geom. 32(3) (2004) 383–400
67. P. Petersen, *Riemannian Geometry*, Graduate Texts in Mathematics, Springer, Berlin Heidelberg New York, 1998
68. J. Rataj, M. Zähle, *Curvatures and currents for unions of sets with positive reach II*, Ann. Glob. Anal. Geom. 20 (2001) 1–21
69. J. Rataj, M. Zähle, *Normal cycles of Lipschitz manifolds by approximation with parallel sets*, Diff. Geom. Appl. 19 (2003) 113–126
70. J. Rataj, M. Zähle, *General normal cycles and Lipschitz manifolds of bounded curvature*, Ann. Glob. Anal. Geom. 27 (2005) 135–156
71. R.T. Rockafellar, *Convex Analysis*, Princeton University Press, Princeton, NJ, 1970
72. R.T. Rockafellar, *Maximal monotone relations and the second derivatives of nonsmooth functions*, Annales de l'institut Henri Poincaré (C) Analyse non linéaire 2(3) (1985) 167–184
73. L.A. Santalo, *Integral geometry and geometric probability*, Addison-Wesley, Reading, MA, 1976
74. R. Schneider, *Convex Bodies: The Brunn–Minkowski Theory*, Cambridge University Press, Cambridge, 1993
75. H.A. Schwarz, *Sur une définition érronée de l'aire d'une surface courbe*, Gesammelte Mathematische Abhandlungen, vol. 1, Springer, Berlin Heidelberg New York, 1890, pp. 309–311
76. M. Spivak, *A Comprehensive Introduction to Differential Geometry*, vol. 1–5, Publish or Perish, Boston, 1971–1975
77. J. Steiner, *Jber Preuss.* Akad. Wiss. (1840) 114–118. In: *Gesammelte Werke*, vol. 2, Chelsea, New York, 1971
78. C. Tricot, *Curves and Fractal Dimension*, Springer, Berlin Heidelberg New York, 1993
79. F. Valentine, *Convex Sets*, McGraw-Hill, New York, 1964
80. S. Wagon, *The Banach–Tarski Paradox*, Cambridge University Press, Cambridge, 1986
81. A. Weinstein, *Lectures on Symplectic Geometry*, Conference Board of the Mathematical Science (CBMS), Regional Conference Series in Mathematics, No. 29, American Mathematical Society, Providence, RI, 1977
82. H. Weyl, *On the volume of tubes*, Am. J. Math. 61 (1939) 461–472
83. P. Wintgen, *Normal Cycle and Integral Curvature for Polyhedra in Riemannian Manifolds*, Differential Geometry (Gy. Soos, J. Szenthe, eds.), North-Holland, Amsterdam, 1982

84. G. Xu, *Convergence analysis of a discretization scheme for Gaussian curvature over triangular surfaces*, Comput. Aided Geom. Des. 23 (2006) 193–207

85. M. Zähle, *Curvature measures and Random sets I*, Math. Nachr. 119 (1984) 327–339

86. M. Zähle, *Curvature measures and Random sets II*, Probab. Theory Rel. Fields 71 (1986) 37–58

87. M. Zähle, *Integral and current representations of Federer's curvature measures*, Arch. Math. (Basel) 46 (1986) 557–567

88. M. Zähle, *Curvatures and currents for union of sets with positive reach*, Geom. Dedicata 23 (1987) 155–171

89. M. Zähle, *Approximation and characterisation of generalized Lipschitz–Killing curvatures*, Ann. Glob. Anal. Geom. 8(3) (1990) 249–260

Index

ε-sample, 256
σ-algebra, 58
k^{th}-mean curvature, 114

affine Grassmann manifold, 107
angular deviation, 200
angular deviation function, 130
approximation of the normal cycle, 200
area formula, 66
area of a surface, 29

ball, 104
basis of the normal cone, 73
Bianchi identity, 102
Borel measure, 60
Borel regular, 60
boundary (of a current), 121

Cauchy formula, 83
change of variable, 64
circumradius, 144
closely inscribed, 136
closely near, 129
coarea formula, 66
connexion form, 103
contact structure, 112
convergence of the normal cycles, 199
convex body, 77
convex hypersurface, 77
convex subset, 77
countably additive, 59
covariant derivative, 113
current, 121
curvature (of a curve), 17
curvature form, 103
curvature measure, 182, 205, 210
curvature tensor, 102

Delaunay triangulation, 253
density, 94
density on a manifold, 98
deviation angle, 130, 146, 200
differential form, 91
dihedral angle, 73
distance function, 47

edge, 71
empty ball property, 254
Euler characteristic, 74
existence of the normal cycle, 196
external dihedral angle, 74

face, 71
fatness, 135, 144
Fenchel-Milnor theorem, 20
flat norm, 123
Frnet equations, 28
Frnet frame, 28
Fubini theorem, 99
fundamental $(N-1)$-form, 214

Gauss curvature, 33, 75
Gauss map, 118
Gauss–Bonnet formula, 104
Gauss-Bonnet theorem, 75
geodesic, 34
geodesic triangle, 34
geometric subset, 197
Grassmann manifold, 105

Hadwiger theorem, 84
Hausdorff distance, 48
Hausdorff measure, 65
homotopy, 124
hypersurface, 113

inscribed, 136
integral (current), 123
integral of a measurable function, 61
internal dihedral angle, 73
invariant form, 189

Kubota formula, 87

Lagrangian submanifold, 112
Lantern of Schwarz, 30
Lebesgue measure, 64
Lebesgue outer measure, 63
Lebesgue theorem, 62
Legendrian, 112
length (of a curve), 13
Levi-Civita connexion, 101
lines of curvature, 33
Liouville form, 111
Lipschitz submanifold, 122
Lipschitz–Killing curvature, 159, 165
local feature size, 55

mass (of a current), 123
mass of the normal cycle, 199
measurable function, 60
measurable space, 58
measure, 57
measure on a manifold, 98
medial axis, 55

normal bundle, 109
normal cone, 73
normal connexion, 116
normal curvature, 117
normal cycle, 193
normalized external dihedral angle, 74
normalized internal dihedral angle, 73

orthogonal projection, 49
outer measure, 57

polyhedron, 71
principal directions, 33
projection map, 49

Quermassintegrale, 154

Radon measure, 67
reach, 52, 177
relative curvature, 133
relative height, 145
restricted Delaunay triangulation, 255
Ricci tensor, 102
Riemannian geometry, 101
Riemannian metric, 101
rightness, 144

sample, 253
scalar curvature, 102
second fundamental form, 33
second fundamental measure, 214
sectional curvature, 102
signed curvature (of a curve), 24
signed measure, 59
simple curve, 20
simple measurable function, 60
simplicial complex, 71
sphere, 104
Steiner formula, 153
submanifold, 109
subset with positive reach, 177
support (of a current), 121
support function, 80
symplectic form, 111

torsion (of a connexion), 101
transversal integral, 85
triangle, 71
triangulation, 144, 253
tube, 48, 165
tubes formula, 168
tubular neighborhood, 48
tubular neighborhood theorem, 110

uniqueness of the normal cycle, 196

valuation, 85
vector valued invariant form, 213
vertex, 71
volume (of a submanifold), 112
Voronoi diagram, 55, 253
Voronoi region, 55

weakly regular, 200
Weingarten tensor, 33

Printing: Krips bv, Meppel, The Netherlands
Binding: Stürtz, Würzburg, Germany